STAAD.Pro 200...
(With U.S. Design Codes)

Munir M. Hamad

ISBN: 1-58503-272-7

SDC
PUBLICATIONS

Schroff Development Corporation

www.schroff.com
www.schroff-europe.com

Version 1.0, 2005

© All rights reserved. No part of this publication may be reproduced or used in any form or by any means – graphic, electronic or mechanical, including photocopying, mimeographing, recording, taping or in information storage and retrieval systems – without the permission of the author or the publisher.

STAAD.*Pro* is a trademark of Research Engineers International (REI), a Bentley Solutions Center (Bentley Systems, Inc.).
Other trademarks are registered trademarks or trademarks of their respective owners.

Tutorial Purpose & Objectives

This tutorial provides an overall look over STAAD.*Pro* 2005. It demonstrates the steps to be followed to produce the structural analysis & design of two types of buildings; concrete and steel. Also the tutorial concentrate over the different results generated from the program, and how to read them, view them, and finally generate the necessary reports from them.

At the completion of this course, the trainee will be able to:

- Understand STAAD.*Pro* way of doing the job
- Creating the geometry using different methods
- Use more advanced technique in creating geometry
- Defining the Cross-Sections of Beams, Columns, Plates
- Defining the Constants, Specifications, and Supports
- Defining the Load Systems
- Analyzing your Model using the appropriate Analysis method
- Reviewing the Analysis Results
- Performing Concrete Design
- Performing Steel Design

Table of Contents

Module 1 Introduction to STAAD.Pro

History of STAAD Software..	1-3
Method of Analysis...	1-5
Three steps to reach your goal..	1-6
Starting STAAD.Pro...	1-8
Creating New File..	1-8
STAAD.*Pro* Screen..	1-11
Opening an Existing File...	1-12
Closing a file..	1-13
Existing STAAD.Pro..	1-14
Saving & Saving As...	1-14
Module Review..	1-15
Module Review Answers...	1-16

Module 2 Geometry

Understanding STAAD.*Pro* Way..	2-3
What are Nodes, Beams, and Plates...	2-4
How things are done in the Input file?...	2-7
Exercise 1...	2-9
Geometry Creation Methods..	2-11
Method 1: Using Structure Wizard...	2-12
Exercise 2...	2-16
Exercise 3...	2-21
Exercise 4...	2-26
Things you can do in Structure Wizard..	2-28
Exercise 5...	2-29
Method 2: Drafting the Geometry using Snap/Grid..	2-32
Exercise 6...	2-36
Viewing..	2-40
Selecting...	2-41
Using Selecting While Viewing 3D Geometry..	2-44
Exercise 7...	2-46
Method 3: Using Copy/Cut with Paste...	2-47
Exercise 8...	2-48
Method 4: Using Spreadsheet (Excel) Copy and Paste..	2-49
Exercise 9...	2-51
Method 5: Using DXF importing file function...	2-53
Exercise 10...	2-56
Workshop 1-A..	2-57
Workshop 1-B..	2-58
Notes..	2-62
Module Review..	2-63
Module Review Answers...	2-64

Module 3 Useful Function to Complete the Geometry

Introduction..	3-3
Translational Repeat...	3-4
Exercise 11..	3-5
Circular Repeat...	3-6
Exercise 12..	3-7
Mirror..	3-8
Exercise 13..	3-9
Rotate..	3-10
Exercise 14..	3-11
Move...	3-12
Insert Node...	3-12
Add Beam Between Mid-Points..	3-14
Add Beam by Perpendicular Intersection...	3-14
Exercise 15..	3-15
Connect Beams along an Axis...	3-17
Intersect Selected Members..	3-17
Exercise 16..	3-18
Cut Section..	3-21
Renumber..	3-23
Exercise 17..	3-24
Delete...	3-26
Undo/Redo...	3-27
Zooming/Panning..	3-27
Dimensioning..	3-29
Pointing to Nodes, Beams, and Plates..	3-30
Global and Local Coordinate System...	3-32
Module Review...	3-35
Module Review Answers..	3-36

Module 4 Properties

Introduction..	4-3
Property Types...	4-3
Type 1: Prismatic..	4-4
Viewing Cross-Section...	4-7
Exercise 18...	4-8
Type 2: Built-In Steel Table...	4-9
Exercise 19...	4-13
Type 3: Thickness..	4-14
General Notes About Property Assigning..	4-15
Workshop 2-A..	4-20
Workshop 2-B..	4-21
Module Review..	4-23
Module Review Answers...	4-24

Module 5 Constants, Supports, and Specifications

Introduction………………………………………………………………………………	5-3
Material Constants………………………………………………………………………	5-3
Exercise 20………………………………………………………………………………	5-6
Geometry Constant………………………………………………………………………	5-8
Exercise 21………………………………………………………………………………	5-10
Supports…………………………………………………………………………………	5-11
How to Assign Supports…………………………………………………………………	5-12
Editing Supports…………………………………………………………………………	5-13
Exercise 22………………………………………………………………………………	5-13
Specifications……………………………………………………………………………	5-14
Exercise 23………………………………………………………………………………	5-18
Workshop 3-A…………………………………………………………………………...	5-19
Workshop 3-B…………………………………………………………………………...	5-20
Module Review…………………………………………………………………………	5-21
Module Review Answers…………………………………………………………….....	5-22

Module 6 Loading

Introduction………………………………………………………………………………	6-3
How to Create Primary Load……………………………………………………………	6-4
Individual Loads: Introduction…………………………………………………………	6-6
Individual Loads: Selfweight……………………………………………………………	6-7
Individual Loads: Members Loads………………………………………………………	6-8
Exercise 24………………………………………………………………………………	6-14
Individual Loads: Area Load……………………………………………………………	6-17
Individual Loads: Floor Load……………………………………………………………	6-18
Individual Loads: Plate Loads……………………………………………………………	6-20
Individual Loads: Node Load……………………………………………………………	6-25
Exercise 25………………………………………………………………………………	6-26
Individual Loads: Viewing & Editing……………………………………………………	6-27
How to Create Manual Combinations……………………………………………………	6-29
How to Create Automatic Combinations…………………………………………………	6-30
Exercise 26………………………………………………………………………………	6-32
Workshop 4-A…………………………………………………………………………...	6-33
Workshop 4-B…………………………………………………………………………...	6-34
Module Review…………………………………………………………………………	6-37
Module Review Answers…………………………………………………………….....	6-38

Module 7 Analysis

Introduction………………………………………………………………………………	7-3
Perform Analysis Command……………………………………………………………	7-3
P-Delta Analysis Command……………………………………………………………	7-6
Non-Linear Analysis Command…………………………………………………………	7-9
The Execution Command………………………………………………………………	7-12
Workshop 5A & 5B………………………………………………………………………	7-16
Module Review…………………………………………………………………………	7-17
Module Review Answers…………………………………………………………….....	7-18

Module 8 Post Processing

Introduction……………………………………………………………………………….…..	8-3
First Step………………………………………………………………………………………..	8-3
Node Displacement……………………………………………………………………………..	8-5
Node Reactions…………………………………………………………………………………	8-10
Beam Forces……………………………………………………………………………………	8-12
Beam Stresses………………………………………………………………………………….	8-16
Beam Graphs…………………………………………………………………………………..	8-19
Plate Contour………………………………………………………………………………….	8-20
Plate Results Along Line……………………………………………………………………..	8-23
Animation…………………………………………………………………………………….	3-26
Reports………………………………………………………………………………………...	8-27
Other Ways: Double-Clicking a Beam……………………………………………………….	8-36
Other Ways: Double-Clicking a Plate………………………………………………………..	8-39
Workshop 6-A & 6-B…………………………………………………………………………	8-40
Module Review………………………………………………………………………………..	8-43
Module Review Answers……………………………………………………………………..	8-44

Module 9 Concrete Design

Introduction…....………………………………………………………………………………	9-3
Modes of Concrete Design…………………………………………………………………….	9-4
Step 1: Job Info………………………………………………………………………………...	9-4
Step 2: Creating Envelopes……………………………………………………………………	9-5
Step 3: Creating Members…………………………………………………………………….	9-6
Step 4: Creating Briefs………………………………………………………………………..	9-8
Step 5: Creating Groups………………………………………………………………………	9-16
Step 6: Design Modes…………………………………………………………………………	9-17
Step 7: Reading Results: Beam Main Layout………………………………………………...	9-20
Step 8: Reading Results: Beam Main Rft…………………………………………………….	9-21
Step 9: Reading Results: Beam Shear Layout………………………………………………..	9-22
Step 10: Reading Results: Beam Shear Rft…………………………………………………...	9-23
Step 11: Reading Results: Beam Drawing……………………………………………………	9-24
Step 12: Reading Results: Column Main Layout……………………………………………..	9-25
Step 13: Reading Results: Column Shear Layout…………………………………………….	9-26
Step 14: Reading Results: Column Results…………………………………………………..	9-27
Step 15: Reading Results: Column Drawing…………………………………………………	9-28
Step 16: Reading Results: Generating Design Reports………………………………………	9-28
Workshop 7A………………………………………………………………………………….	9-32
Module Review………………………………………………………………………………..	9-35
Module Review Answers……………………………………………………………………..	9-36

Module 10 Steel Design

Introduction..	10-3
Step 1: Load Envelope Setup..	10-4
Step 2: Member Setup...	10-5
Step 3: Change the Restraints...	10-6
Step 4: Creating Briefs..	10-9
Step 5: Creating Design Groups...	10-14
Steel Design Commands in STAAD.Pro..	10-15
Workshop 7-B..	10-18
Notes...	10-21
Notes...	10-22
Module Review..	10-23
Module Review Answers...	10-24

Preface

- STAAD.*Pro* is a tool for structural engineers.
- This tutorial is meant for the new users of STAAD.*Pro* 2005, whom didn't work before on STAAD.Pro, but possesses reasonable experience of Windows OS.
- This tutorial is NOT a replacement of the manuals of STAAD.Pro; on the contrary, we encourage all the readers to read the manuals thoroughly. We consider this tutorial as the first step for the beginners, which after finishing it, and with the help of STAAD.*Pro* manuals along with the Help system the user will be able to master all of the other features of STAAD.Pro.
- This tutorial's main objective is to go with the novice user step-by-step starting from creation of the geometry up until performing concrete and steel design.
- The user should have enough experience in the manual methods, as neither STAAD.*Pro* nor this tutorial will teach any manual structural methods.
- This tutorial can be used as *instructor-led* tutorial, or *teach-your-self* tutorial:
  - As for the first option the estimated time would be 3 days, 8 hours a day.
  - As for the second option, the reader can take it up to his/her convenience.
- There are 26 exercises to be solved, each after certain topic discussed. The main reason of these exercises is to let the user practically go through the procedure, rather than just reading about it.
- Also, there are 14 workshops; 7 for concrete, and 7 for steel. It is preferable to go through all of them, so the reader will be exposed to all the functions of STAAD.Pro.
- This tutorial will cover the basic and intermediate levels of knowledge in STAAD.Pro.
- This tutorial is covering STAAD.*Pro* 2005, and it is designed for the people who use the American Codes for both Concrete Design, and Steel Design using Metric units.

Notes:

Module 1:

Introduction to STAAD.*Pro*

This module contains:

- History of STAAD Software
- Method of Analysis
- Three steps to reach your goal
- Filing System of STAAD.*Pro*

History of STAAD Software

- STAAD stands for **ST**ructural **A**nalysis **A**nd **D**esign. It is one of the first software applications in the world made for the purpose of helping the structural engineers to automate their work, to eliminate the tedious and lengthy procedures of the manual methods. Its history is as follows:

STAAD-III for DOS
- STAAD first versions were built for DOS Operating System, and it was non-graphical software. The user should first undergo a lengthy reading to understand the syntax of STAAD language of commands in order to create the input file, then will send this file to the analysis and design engine to execute it. Text output will be produced accordingly.
- With time, STAAD.Progress to create it's own graphical environment, this was a major change for the STAAD users, as they were able to build their input file without the need to understand the syntax of STAAD language of commands but still the interface was not user friendly.

STAAD-III for Windows
- Research Engineers, Inc. (REI) worked in two parallel lines to provide STAAD for Windows:
  - They made *not-really-Windows* application which works under Windows environment. The new software looked like STAAD-III for DOS, so all of what you have to do is to switch to Windows and start working, no need for any new experiences.

  - The second track was REI & QSE merged. QSE has a very real-Windows interface, but lacks the power of STAAD engines in both analysis and design areas, plus the superiority of STAAD multi-coded design engines, which supports almost all of the famous codes in the world. REI and QSE joined forces to produce STAAD.*Pro*, which was a milestone in both STAAD history and structural analysis and design software industry.

STAAD.Pro
- STAAD.*Pro* was born giant. It was mixture of the expertise of two long experienced companies.
- STAAD.*Pro* introduced a really good-looking interface which actually utilized all the exceptional features of Windows 95/98/2000/ME/XP (Each STAAD.*Pro* was working respectively under the Windows available at the time of releasing the software to the markets). This new interface empower the user of STAAD.*Pro* to accomplish the most complicated structural problems in short time, without scarifying the accuracy and the comprehensive nature of the results.
- STAAD.*Pro* with its new features surpassed its predecessors, and compotators with its data sharing capabilities with other major software like AutoCAD, and MS Excel.
- The results generation was yet a new feature that you can depend on STAAD.*Pro* to do for you, now, STAAD.*Pro* can generate handsome reports of the inputs and the outputs with the usage of graphical results embedded within, which can be considered as final document presented to the client.
- The concrete and steel design were among the things that undergone a face-lift, specially the concrete design, as REI created a new module specially to tackle this issue. This new module is easy, and straightforward procedure making the concrete design and results generation a matter of seconds ahead of the user.

Method of Analysis

- One of the most famous analysis methods to analyze continuous beams is "Moment Distribution Method", which is based on the concept of transferring the loads on the beams to the supports at their ends.
- Each support will take portion of the load according to its *K*; *K* is the stiffness factor, which equals *EI/L*. As you can see *E*, and *L* is constant per span, the only variable here is *I*; moment of inertia. *I* depends on the cross section of the member. So, if you want to use this analysis method, you have to assume a cross section for the spans of the continuous beam.
- If you want to use this method to analyze a simple frame, it will work, but it will not be simple, and if you want to make the frame a little bit more complicated (simple 3D frame) this method falls short to accomplish the same mission.
- Hence, a new more sophisticated method emerged, which depends fully on matrices, this method called "Stiffness Matrix Method", the main formula of this method is:

$$[P] = [K] \times [\Delta]$$

- The 3 matrices are as follows:
  - [P], is the force matrix, which includes the forces acting on the whole structure, and the reactions at the supports. This matrix is partially known, as the acting forces on the structures are already known from the different codes, like Dead Load, Live Load, Wind Load, etc., but the reactions are unknown.
  - [K], is the stiffness factor matrix. *K=EI/L*, and all of these data either known or assumed. So this matrix is fully known.
  - [Δ], is the displacement matrix. The displacements of supports are either all zeros (fixed support) or partially zeros (other supports), but the displacements of other nodes are unknown. So this matrix is partially known.

- With these three matrices presented as discussed above, the method will solve the system with ordinary matrix methods to get the unknowns. If we solved for the unknowns, the reactions will be known, hence shear and moment diagrams can be generated, and the displacement of the different nodes will be known, so the displacement and deflection shapes can be generated.
- This method was very hard to be calculated by hand as it needs more time than other methods, so, it was put on the shelves, up until the emergence of computers. The different programming languages revive the possibility to utilize this method, as the program will do all the tedious and lengthy procedures to solve for this system of matrices, therefore, structural software adopted it as the method of analysis. STAAD was one of the first to do that.

Three steps to reach your goal

- There are three steps to reach to your goal:
  - Prepare your input file.
  - Send your input file to the analysis/design engine.
  - Read the results and verify them.

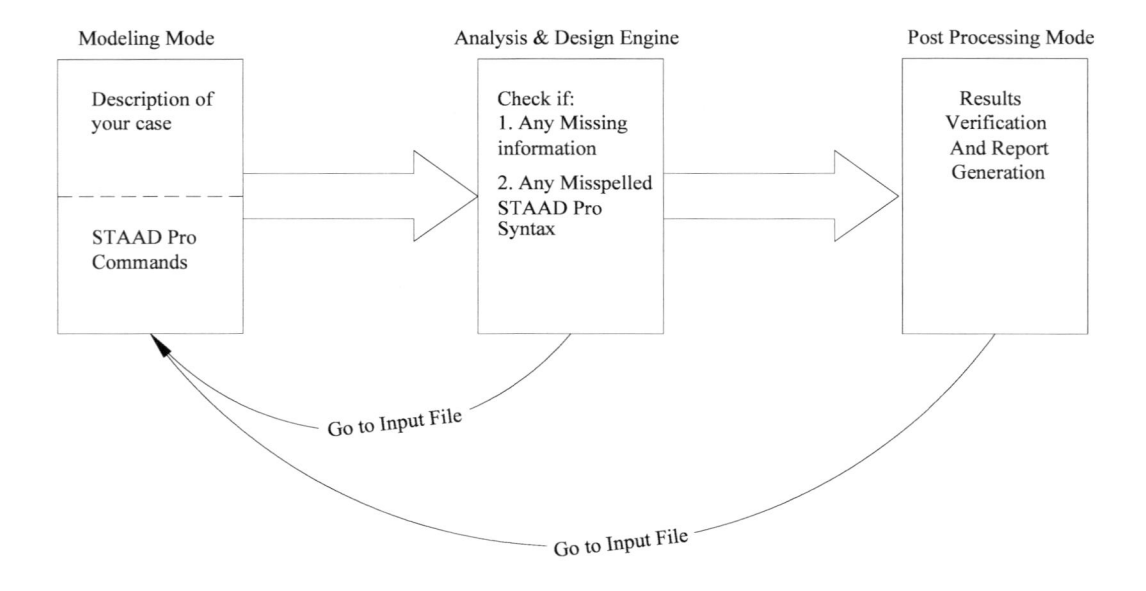

Modeling Mode

Description of your case

STAAD Pro Commands

Analysis & Design Engine

Check if:
1. Any Missing information
2. Any Misspelled STAAD Pro Syntax

Post Processing Mode

Results Verification And Report Generation

Go to Input File

Go to Input File

Input file
- Creating input file takes place in the **Modeling Mode**. It is your first step in working in STAAD.*Pro*. *What is input file?* Input file is the place you describe your case; what do you have? And what do you want? We can cut the input file into two parts:
  - In the first part you will describe your structure. This includes the geometry, the cross sections, the material and geometric constants, the support conditions, and finally the loading system.
  - The second part *may* contain the analysis command, and printing commands.

Send your input file to the analysis and design engine
- Just like any programming language compiler, STAAD.*Pro* analysis and design engine, will start reading the input file from left to right, and from top to bottom. The engine will mainly check for two things:
  - Making sure that the user used the syntax of STAAD.*Pro* commands, or else the engine will produce an error message.
  - Making sure that all the data needed to form a stable structure exists in the input file, or else, the engine will produce an error message.
- If these two things are correct, STAAD will take the values mentioned in the input file (without verification) and produce the output files.
- As a rule of thumb, generating the output files doesn't mean that results are correct! The concept of GIGO (Garbage In Garbage Out) applies. Based on this concept, don't take the results generated by STAAD.*Pro* as final, but verify each piece of the output data, to make sure that your input data was correct.

Read results, and verify them
- Reading output takes place in **Post Processing** Mode. It includes:
  - Seeing the results as tables and/or as graphical output.
  - Changing the scale of each graphical output to visualize the correct shapes, and showing values, or hiding them.
- After reading and verifying your results you may decide to go back to your Modeling Mode to alter your input file, for either to correct the input file, or to change some values to examine different results. The input file always has extension of **STD**.

Starting STAAD.*Pro*

- There are two possible ways to start STAAD.*Pro*:
    - Go to Start/All Programs/STAAD.*Pro* 2005/STAAD.*Pro*.
    - Double-click the shortcut on the Windows Desktop.

Creating new file

- Creating new file in STAAD.*Pro* can be done in two different ways:
    - Once you started the software.
    - The software is already running and you want to create new file, select **File/New**, or click the **New Structure** button in the **File** toolbar. In both ways, the same dialog box will be displayed.

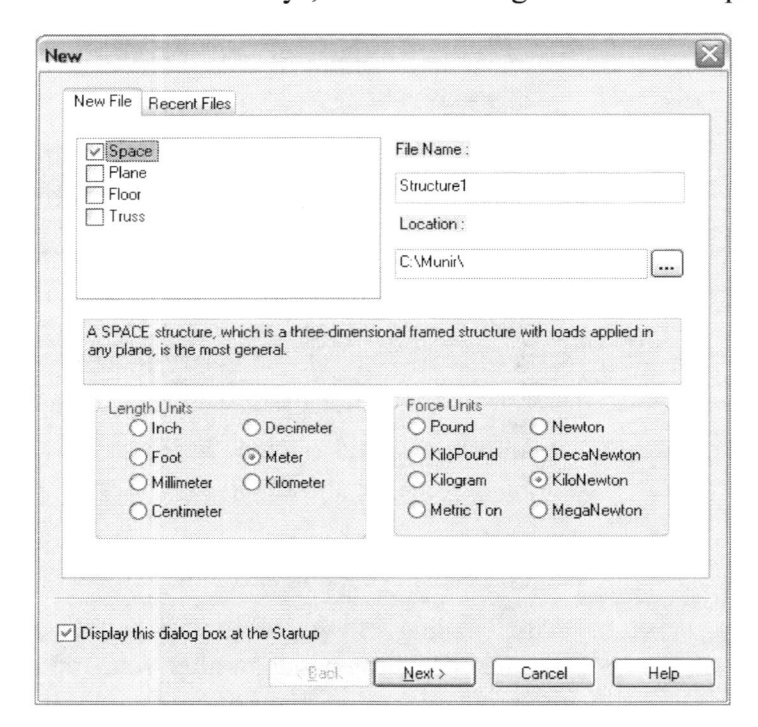

- STAAD.*Pro* can deal with single file at a time, so, if you attempt to create a new file, while another file is opened, STAAD.*Pro* will close it right away. The parts of this dialog box are:

File Name
- Specify the name of the new file (no need to type .STD, STAAD will do that for you); file names in STAAD.*Pro* can take long file names.

Location
- Specify where you will save this file in your local hard drives, or any network hard drive, and then specify the folder name (sub-directory) (example F:\SPRO2005\ STAAD\Examp), To change these settings, simply click the three dots button, and the following dialog will appear:

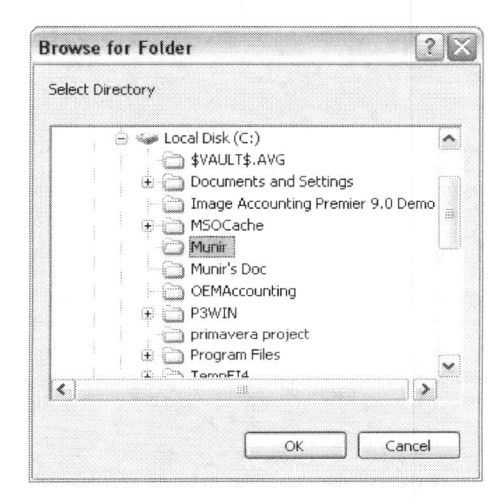

Type of Structure
- STAAD.*Pro* provides 4 different structure types.
  - **Space**: Three-dimensional framed structure with loads applied in any plane (The most general).
  - **Plane**: Two-dimensional structure framed in the X-Y plane with loads in the same plane.
  - **Floor**: Two, or three-dimensional structure having no horizontal (global X or Z) movement of the structure (FX, FZ & MY, are restrained at every joint).
  - **Truss**: Any structure consists of truss members only, which can have only axial member forces and no bending in the members.

Length, and Force Units

- When you install the software at your hard drive, the installation software will ask you to specify what is your default unit system, English (ft, inch, kips) or Metric (m, mm, KN). For this tutorial we chose Metric, hence the default Length, and Force Units are Meter, and Kilo Newton respectively.
- This will be *to-start-with* units, and not the only units you can use while you are creating the input file. As a user you have the ability to change the units at any point to whatever desired units (STAAD internally will make the necessary conversion).
- When you are done click **Next** in order to proceed. The following dialog box will be displayed:

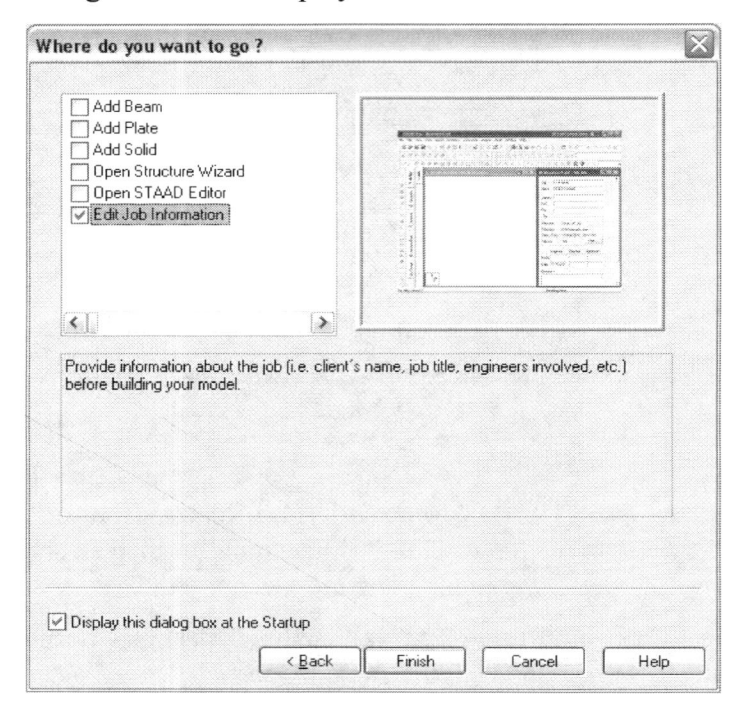

- The only purpose of this dialog box is to ask the user what is the first step to be done in creating the input file? We will choose the last option: **Edit Job Info**, as all of the other options will be discussed whilst we are in the **Geometry** part of the input file.
- To finish creating a file in STAAD.Pro, click **Finish**.

STAAD.*Pro* Screen

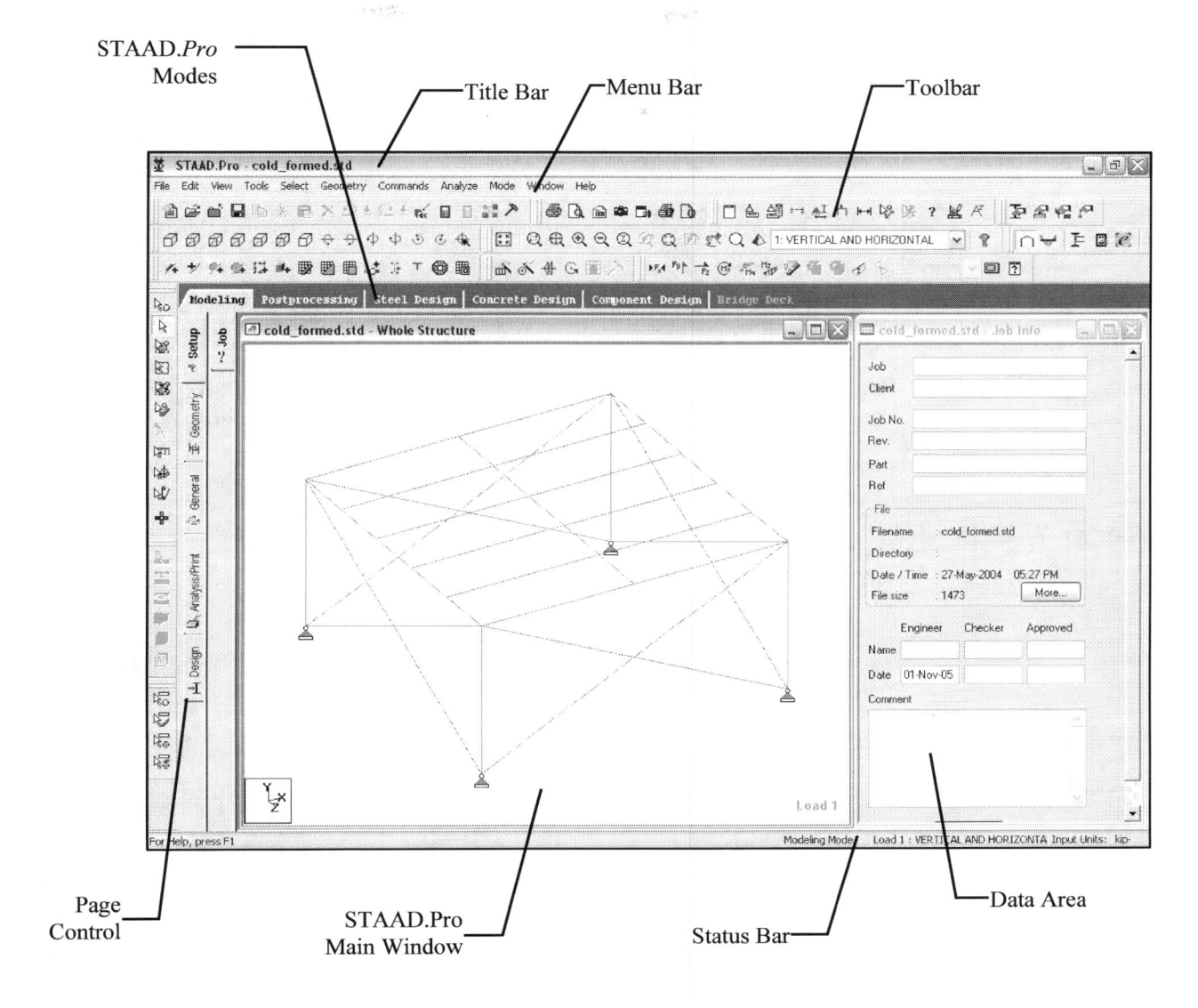

Labeled screenshot of the STAAD.Pro interface with callouts: STAAD.*Pro* Modes, Title Bar, Menu Bar, Toolbar, Page Control, STAAD.Pro Main Window, Status Bar, Data Area.

Notes on Page Control & Data Area

- Page Control is another way (after menus, and toolbars) to execute commands in STAAD.Pro.
- Page Controls are:
  - The tabs that appear at the left of the main window.
  - Each Page Control has its own sub-pages.
  - Each Page Control has its own function, which will help the user to accomplish one of the tasks required.

- The sequence of the Page Control is meant to be like this. If you follow the pages and sub-pages in this sequence, you will fulfill the task of creating a complete input file, without missing any essential detail. This method helps doing your job, fast and accurate.
- Page Control is meaningless without the linked Data Area (the part at the right of the main window). Data Area will give two things:
  - It will show the relevant data of your structure related to the current Page Control (e.g. If you are in the Geometry Page Control, Data Area will show Node Coordinates and Beams Incidences)
  - It will show relevant buttons (which represents commands) to add/edit commands related to the current Page Control.
- In this tutorial will concentrate more on toolbars, and Page Control & Data Area in issuing STAAD.*Pro* commands, and utilities.

Open an existing file

- Opening an existing file in STAAD.*Pro* can take place in three different ways:
  - While you are starting STAAD.Pro, select **Recent Files**, the following dialog box will appear:

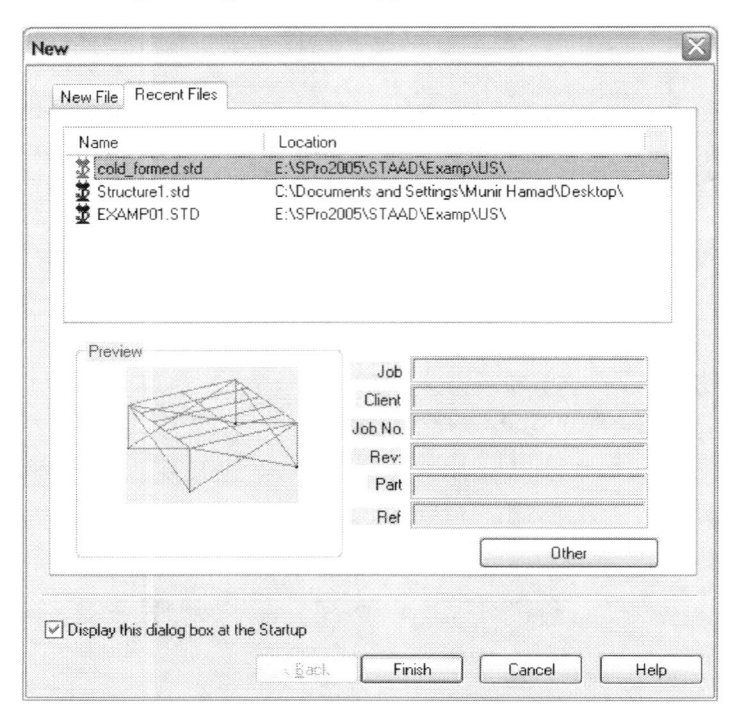

Module 1: Introduction to STAAD.*Pro*

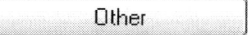

[Other]
- If your file is not among the files listed, simply click **Other** button, select the desired drive, and folder, then select STAAD.*Pro* file, and click **Open**. Check the below dialog box:

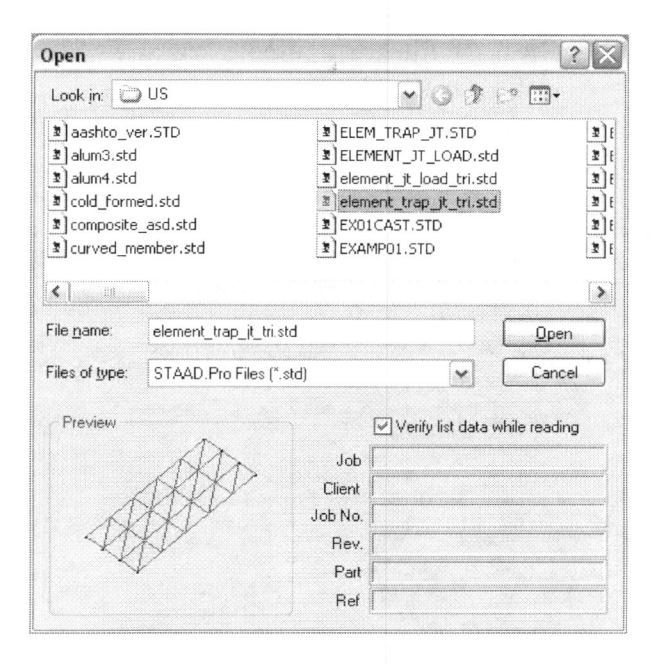

- The software is already running, and you want to open another file, select **File/Open**, or click **Open Structure** button from the **File** toolbar, as a result the same dialog box will appear, do as listed above.

Closing a file

- You can close file in STAAD.*Pro* without existing STAAD.Pro. Select **File/Close**, or click **Close Structure** button from **File** toolbar.

Note
- When you are closing a file without saving it, STAAD.*Pro* will give you warning: *this file will be closed without saving the changes*, so, you will have the choice either:
  - Saving the file now.
  - Close without saving the file.
  - Canceling the operation of closing the file

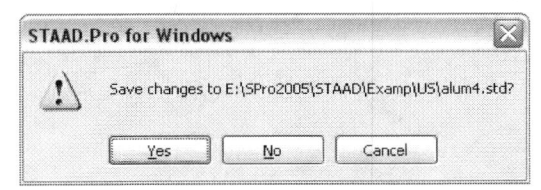

1-13

Exiting STAAD.*Pro*

- To exit STAAD.*Pro* select **File/Exit** and STAAD.*Pro* will close the current file, and exit the software. The only difference between closing a file and exiting STAAD.*Pro* is the closing of the software, and the rest is the same.

Saving and Saving As

- To save the current file, you can select **File/Save**, or click the **Save** button in the File toolbar.
- To save the current file under a new name, simply select **File/Save As**, the below dialog box will be displayed.

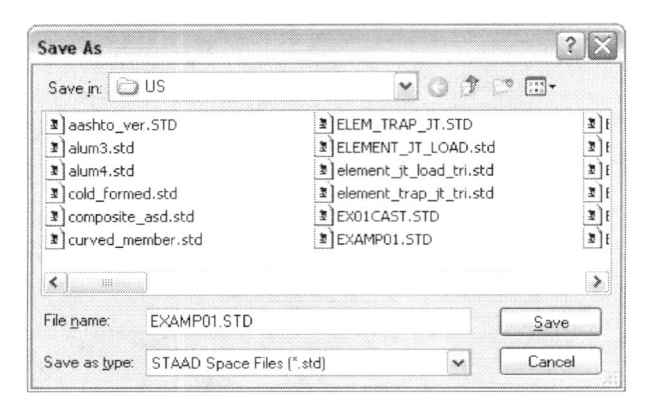

- First select the desired drive, and folder. Then, type in the file name, leave the file type to be STAAD Space File (*.std), click **Save**.

Module Review

1. The new generation of STAAD is:

    a. STAAD-III for DOS

    b. STAAD-III for Windows

    c. STAAD.*Pro* for DOS

    d. STAAD.*Pro*

2. You are NOT obliged to input member cross section if you want to deal with the Stiffness Matrix Method:

    a. True

    b. False

3. Page Control and _____ are linked together.

4. Default Units are specified in the Installation process:

    a. True

    b. False

5. STAAD can deal with:

    a. 2 files at a time.

    b. 4 files at a time.

    c. 1 file at a time.

    d. All of the above.

Module Review Answers

1. d
2. b
3. Data Area
4. a
5. c

Module 2:

Geometry

This module contains:

- Understanding STAAD.*Pro* way
- What are Nodes, Beams, and Plates?
- How things are done in the input file?
- Using Structure Wizard to create Geometry
- Using Drafting to create Geometry
- Using Copy/Cut with Paste to create Geometry
- Using Spreadsheet to create Geometry
- Using DXF importing to create Geometry

Understanding STAAD.*Pro* way

- In order to build up a good input file we have to understand STAAD.*Pro* way. This procedure will enable us to:
  - Organize our thoughts.
  - Put each step in its right position, not before, and not after.
  - Make sure that all of the STAAD.*Pro* commands are present in the input file (none of them is overlooked).
  - Provide us with speedy and guaranteed way to create the input file.
  - Avoid error messages.

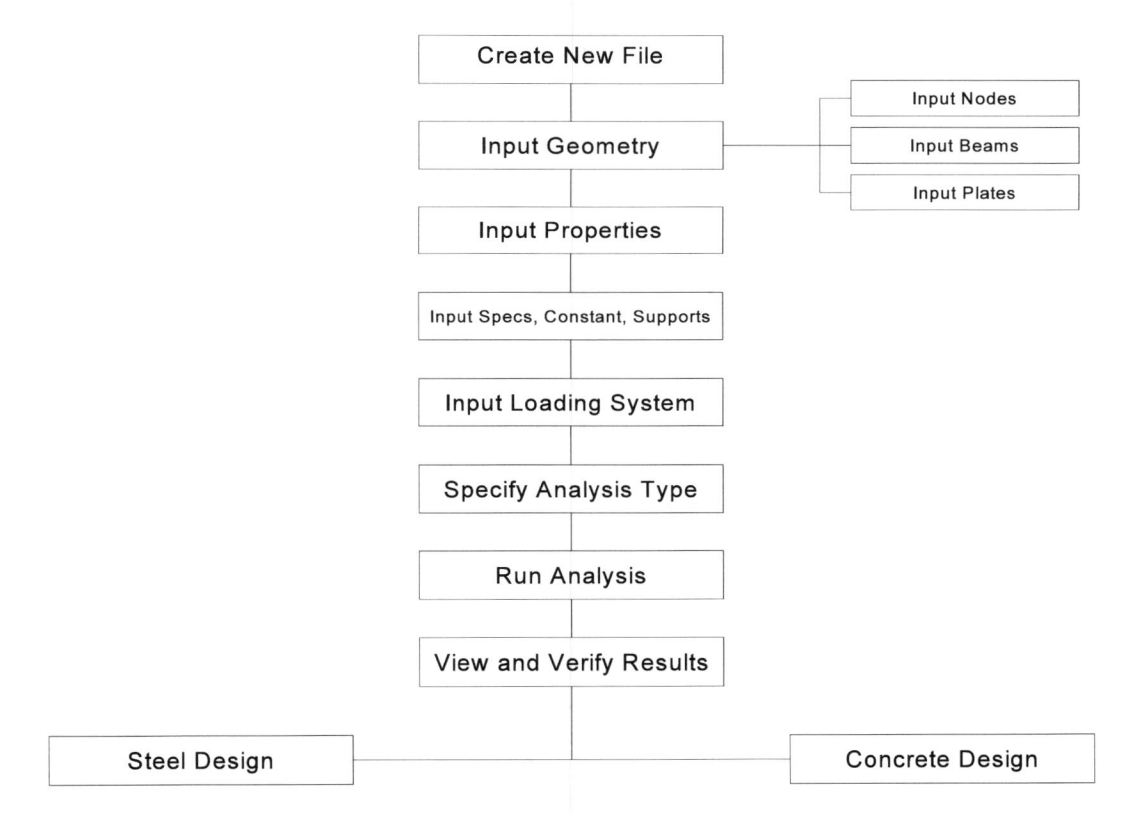

```
Create New File
      ↓
Input Geometry ──────→ Input Nodes
      ↓                Input Beams
      ↓                Input Plates
Input Properties
      ↓
Input Specs, Constant, Supports
      ↓
Input Loading System
      ↓
Specify Analysis Type
      ↓
Run Analysis
      ↓
View and Verify Results
   ↙        ↘
Steel Design   Concrete Design
```

- If this path was followed sincerely, the creation time of your input file will be cut by 50%, that's why this will be our procedure through out this tutorial.
- As you can see from the above flow chart, the second step after creation of a new file is to input the Geometry of your structure. Geometry is the subject of this module, so; what exactly STAAD.Pro means by Geometry?
- Geometry is the "skeleton of your structure", or, in other words Geometry is "the members (beams and columns), and the plates (slabs, walls, and raft foundations)". Through the information you will provide in this part of the input file, STAAD.Pro will understand the following:
  - In which plane (X-Y, Y-Z, X-Z, or any other custom planes) each member and plate is defined?
  - What is the dimension of each member, and plate?
  - What is angle of each member in the space?
  - How members are connected to each other, and how they are connected to the plates?

What are Nodes, Beams, and Plates?

Node
- **Node** in STAAD.Pro means; Stiffed joint with 6 reactions.
- It is located at each end of Beam, and each corner of Plate. Nodes considered the essence of the Geometry of any structure in STAAD.Pro. Each Node will hold the following information:
  - Node Number.
  - Node Coordinate in XYZ space.

Beam
- Beam in STAAD.Pro means; any member in the structure.
- It can be beam, column, bracing member, or truss member.
- Beams are actually defined based on the Nodes at their ends. Each Beam will hold the following information:
  - Beam Number.
  - The Node numbers at its ends.

Example   Node # 1 Coordinate is 0,0,0
Node # 2 Coordinate is 0,2,0
Node # 3 Coordinate is 2,2,0
Node # 4 Coordinate is 2,0,0

Beam # 1 Between Node 1 and 2
Beam # 2 Between Node 2 and 3
Beam # 3 Between Node 3 and 4

Note   ■ Z coordinate in all coordinates is 0; hence this structure lies in the X-Y plane. See the figure below.

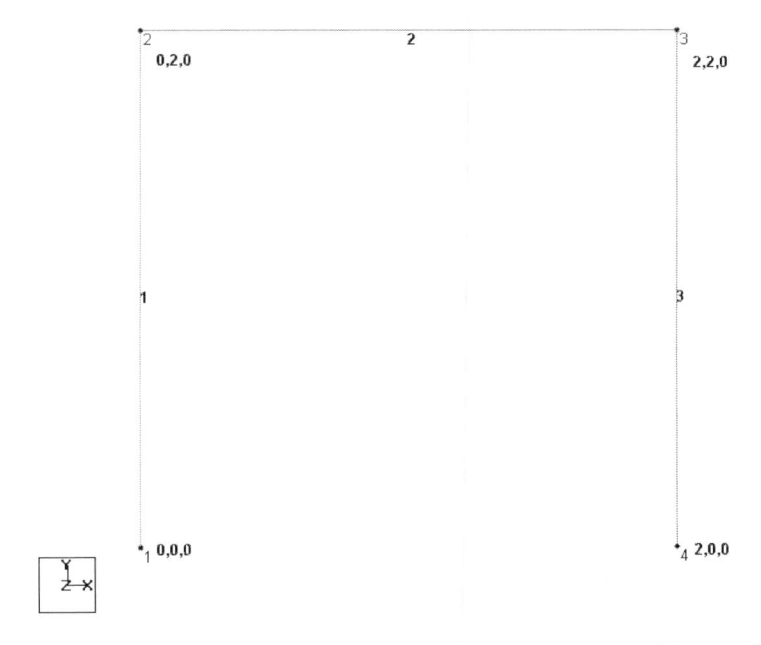

Plate   ■ Plate in STAAD means; a thin shell with multi-nodded shape starting from 3 nodes, and more.
■ It can be anything of slab, wall, or raft foundation. Each Plate will hold the following information:
• Plate Number.
• Node Numbers at each corner of it.

Example   Node # 3 Coordinate is 0,2,0
Node # 4 Coordinate is 2,2,0
Node # 8 Coordinate is 2,2,2
Node # 7 Coordinate is 0,2,2

Plate # 9 Between Nodes 3, 4, 8, 7

Note ■ Y-coordinate is the above four Nodes is constant (namely; 2), and X, and Z is variable, hence the plate is located in the X-Z plane. See the figure below.

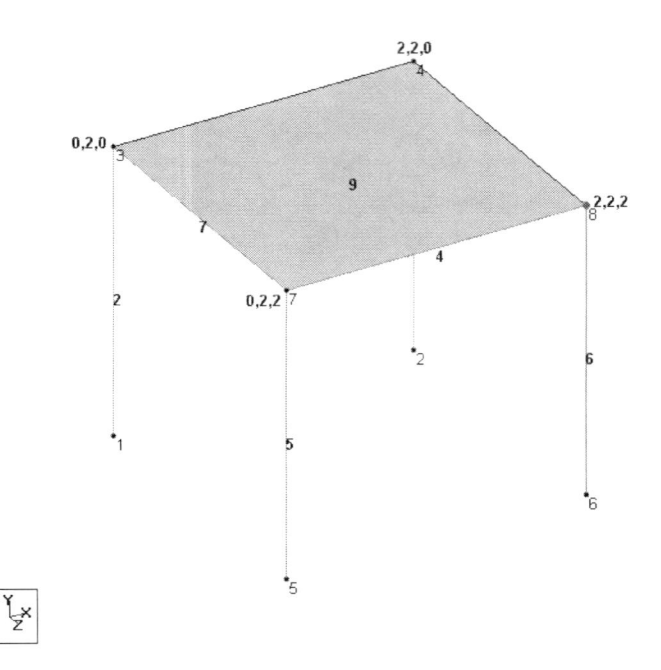

How things are done in the Input file?

- STAAD.*Pro* will create the contents of the input file concerning geometry, and hence it will number all the Nodes, Beams, and Plates. But how they are created?
- STAAD has it's own syntax of creating the input file, goes like this:
  JOINT COORDINATES
  1 0 0 0; 2 0 2 0; 3 2 2 0; 4 2 0 0; 5 0 0 2; 6 0 2 2; 7 2 2 2
  8 2 0 2;
  MEMBER INCIDENCES
  1 1 2; 2 2 3; 3 3 4; 4 2 6; 5 3 7; 6 5 6; 7 6 7; 8 7 8
  ELEMENT INCIDENCES
  9 3 4 8 7;

- Did you understand what each number means in the three sections?

Explanation
- In the Joint Coordinate section the following applies:
  - The first number is the Node Number.
  - The three other digits are the coordinates of the Node.
  - Semi-colon is used to separate each Node data from the other.
  - If one line in the editor is not enough, you can use the next line without semi-colon.

- In the Member Incidences section the following applies:
  - The first number is the Beam Number.
  - The other two digits represent the Node numbers at its ends.
  - Semi-colon is used to separate each Beam data from the other.
  - If one line in the editor is not enough, you can use the next line without semi-colon.

- In the Element Incidences section the following applies:
  - The first number is the Plate Number.
  - The other four digits represent the node numbers at its corners (this example contains a 4-noded plate, hence we used four digits, but this number can be reduced to 3, or increased to more than 4).
  - Semi-colon is used to separate each Plate data from the other.
  - If one line in the editor is not enough, you can use the next line without semi-colon.

Clarification
- We have to clarify some naming convention problems, which may confuse the reader of this tutorial. STAAD.Pro uses the following terms in the *graphical part* of Modeling Mode:
  - Node
  - Beam
  - Plate
- On the other hand, STAAD.Pro uses the following naming convention for the same in the *text editor*:
  - Node becomes Joint.
  - Beam becomes Member
  - Plate becomes Element
- This confusion is a result of joining QSE and STAAD-III for Windows; accordingly the first set of naming is used by QSE, whereas the second set is used by STAAD-III for Windows. After the emergence of the two software packages, each software package kept its own naming convention. Within our discussion we will use the first naming convention (namely; Node, Beam, and Plate).
- Another naming convention, which may create confusion, is when STAAD.Pro calls Beam for both beams and columns. That is correct almost in all of the places of the software except in the concrete design module, when the software distinguish beams from columns. So, if we want to select a column in STAAD.Pro, and you read in this tutorial click on the Beams Cursor, don't get confused!

Practicing Geometry Creation

Exercise 1

1. Using the Structure in the below figure, do the following:

    a. Number all Nodes starting from Node 1.

    b. Number all Beams.

    c. Number all Plates.

    d. Write on the figure the coordinate of each node (check the XYZ icon at the lower left corner of the figure).

    e. Write the three sections of Joint Coordinates, Member Incidences, and Element Incidences.

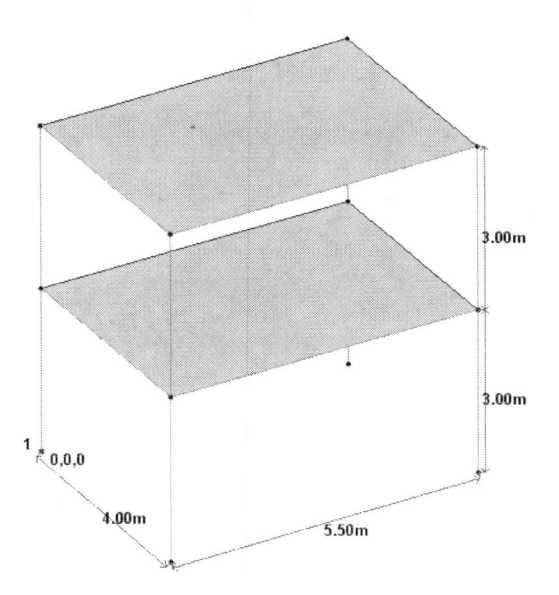

Solution of Exercise 1

Geometry Creation Methods

- STAAD.*Pro* comes with intelligent, accurate, speedy, error-free, and graphical methods to accomplish the creation of Geometry. These are:
  - Using Structure Wizard.
  - Drafting the geometry using the Snap/Grid.
  - Using Copy/Cut, with Paste.
  - Using Spreadsheet (namely; Excel) Copy and Paste.
  - Using DXF importing file function.
- Each one of these 5 methods (by itself) can help the user reduce the time of creating the geometry needed. Alternatively, user can't accomplish the whole process of creating geometry with any of these methods alone; instead, user will need more functions to make necessary modification on the geometry to render the final shape. These functions will be the subject of Module 3.

Method 1: Using Structure Wizard

- Structure Wizard is a library of pre-defined structural shapes allows the user to create a full structure by answering simple questions about the dimensions of members in each axis. From the menus select **Geometry/Run Structure Wizard**; the following window will appear:

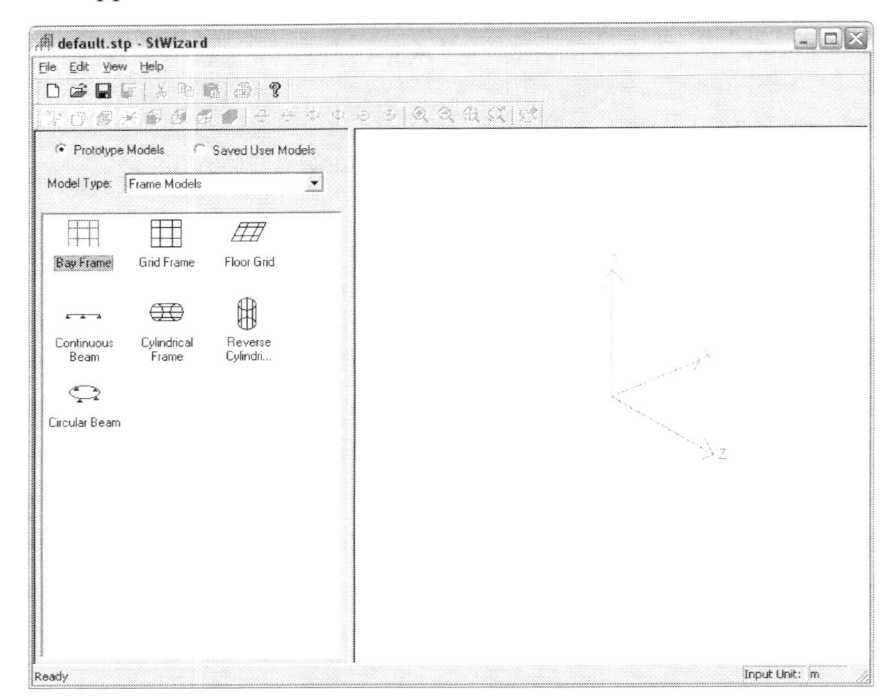

- There is a general method to utilize Structure Wizard effectively for all types of the structure:
- From the left part, select the **Model Type**, there are 7 of them:
    - Truss Models
    - Surface/Plate Models
    - Composite Models
    - VBA-Macro Models
    - Frame Models
    - Solid Models
    - Import CAD Models

Module 2: Geometry

- In the lower screen beneath the **Model Type**, STAAD.*Pro* will show the available structures in this type, as an example, in the Frame Models, the following structures are available:

  - Bay Frame.
  - Floor Grid
  - Cylindrical Frame
  - Circular Frame
  - Grid Frame
  - Continuous Beam
  - Reverse Cylindrical Frame.

- Double-click on the desired structure.
- The **Select Parameters** dialog box will appear. This dialog box will show different type of parameters for each structure (we will discuss each case by itself).
- Fill in the data, and click **Apply**.
- Select **Edit/Add Paste Model in STAAD.*Pro*** from menus, or click **Transfer Model** icon from the toolbar.
- The confirmation message will be shown, to confirm that the user wants really to transfer the model created in Structure Wizard to STAAD.*Pro* window.

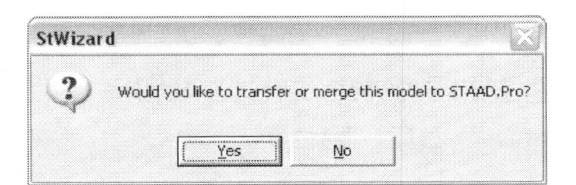

- Click Yes. Now STAAD.*Pro* will ask the user to specify the pasting point in the XYZ space, as shown below. As you can see the default pasting point is 0,0,0 which is the best point if there is no other structure in the STAAD.*Pro* window, but if there is a structure, a different point will be entered (check Reference Point).

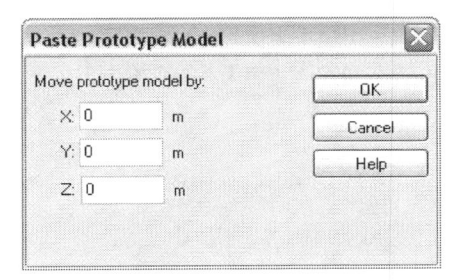

2-13

- Click **OK**, the model created in the Structure Wizard will be pasted in STAAD.*Pro* window as required.
- This is a general method, which is applicable to all types of the structures embedded in the library. Now we will discuss each type by it self.

Frame Models / Bay Frame

- **Bay Frame** is any 3D structure frame consists of beams and columns.
- After you start Structure Wizard, select from the model pop-up list **Frame Models**, the following structures will be shown.

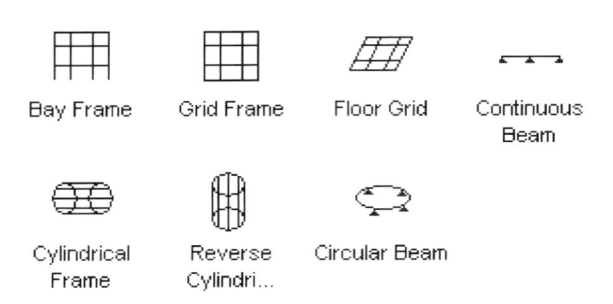

[Icons: Bay Frame, Grid Frame, Floor Grid, Continuous Beam, Cylindrical Frame, Reverse Cylindri..., Circular Beam]

- Double click on the Bay Frame icon to setup the dimensions. The following dialog box will be displayed.

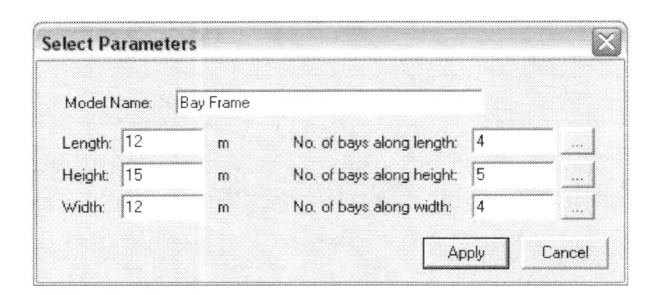

[Dialog: Select Parameters
Model Name: Bay Frame
Length: 12 m   No. of bays along length: 4
Height: 15 m   No. of bays along height: 5
Width: 12 m   No. of bays along width: 4
Apply   Cancel]

- Now specify the following inputs:
  - The Length (Length is in X direction).
  - The Height (Height is in Y direction).
  - The Width (Width is in Z direction).
  - Number of bays along length.
  - Number of bays along height.
  - Number of bays along width.
  - Click Apply.

Note
- All the numbers should be positive.
- If you don't want one of the dimensions, simply set it to be zero, the structure will become two-dimensional.
- You should input the total dimension in each side; that is the total Length, total Height, and the total Width.
- Bay means span.
- If you have a Length of 12 m, and Number of Bays of 2, by default each Bay will be 6 m long.
- If the spans are not equally spaced, click the button with the three dots (to the right of **Number of bays** field) to set the distances of each span. Check the dialog box below.

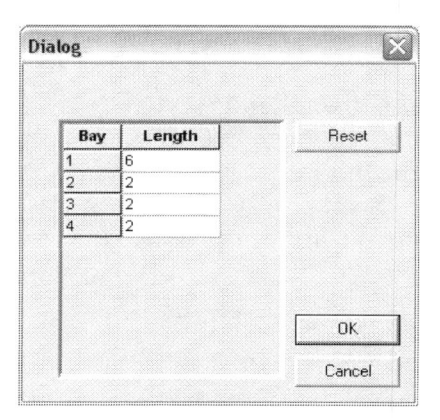

Note
- Always consider the lengths from left-to-right, from bottom-to-top, and from behind-to-front.
- Make sure that the sum of the spans equals the dimension, as STAAD.*Pro* will produce an error message warning you to correct this error, check the figure below. Click **OK** to accept the numbers.

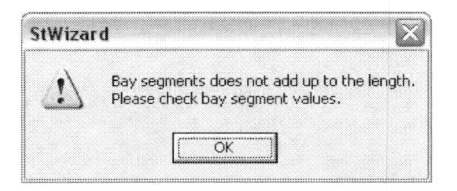

Using Structure Wizard to Create Bay Frame

Exercise 2

1. Start STAAD.*Pro*.
2. Create a new file using the following data:
    a. Space
    b. Units: Meter, and KiloNewton
    c. Click Edit Job Info
3. Using Structure Wizard, try to create the structure shown below:

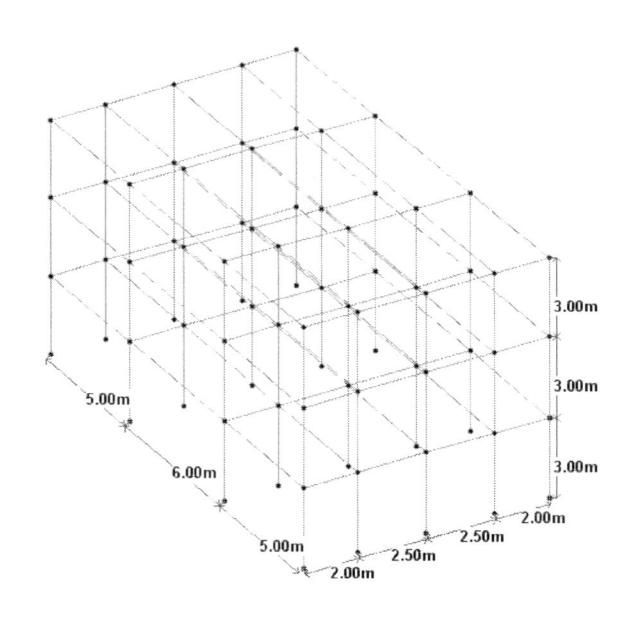

4. Keep the file open; you will need it in the next exercise.

Frame Models / Grid Frame
- **Grid Frame** is just like Bay Frame with one exception, it creates ground beams in the X-Z plane of the structure.
- Check the illustration below to compare between Bay Frame and Grid Frame.

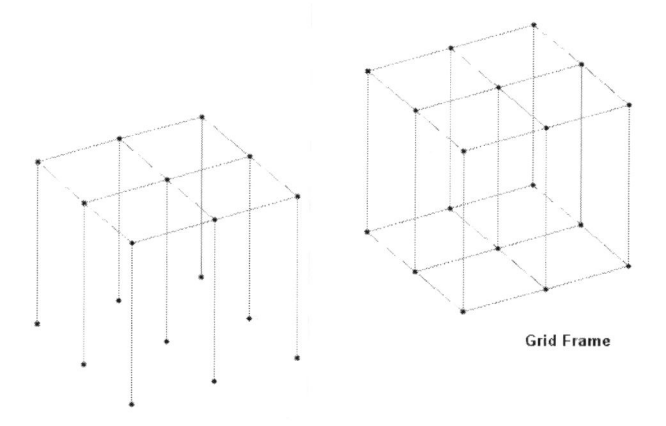

Grid Frame

Frame Models / Floor Grid
- **Floor Grid** is two-dimensional structure is the X-Z plane only.
- The purpose is to create a mesh of beams in the X and Z direction. Double-click on the **Floor Grid** icon, the following dialog box will be shown. Note that the Height (Y-Axis) is grayed out:

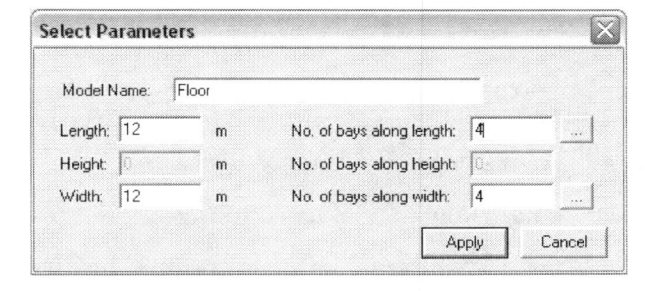

Frame Models / Continuous Beam
- Continuous Beam is one-dimensional structure in the X direction on.
- Double-click **Continuous Beam** icon, the following dialog box will be shown. Note that Height (Y Axis) and Width (Z Axis) are grayed out; hence they are not available for editing.

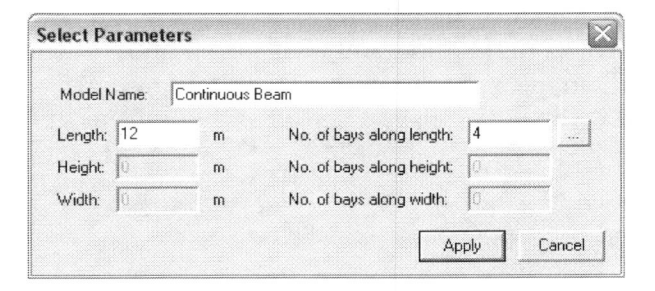

Truss Models / All types

- From the Model pop-up list, select **Truss Models**
- The following structures will be shown.

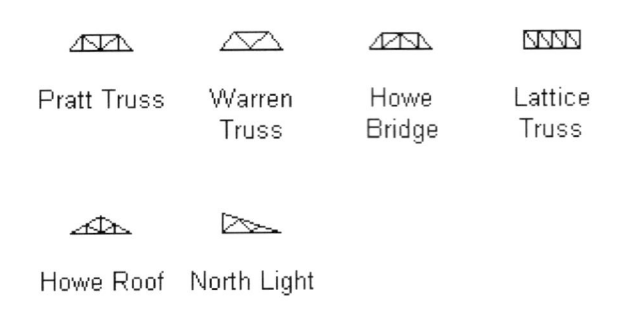

  Pratt Truss   Warren Truss   Howe Bridge   Lattice Truss

  Howe Roof   North Light

- If you double-click on any of the icons you will get the same dialog box for all six shapes, as shown in the dialog box below.

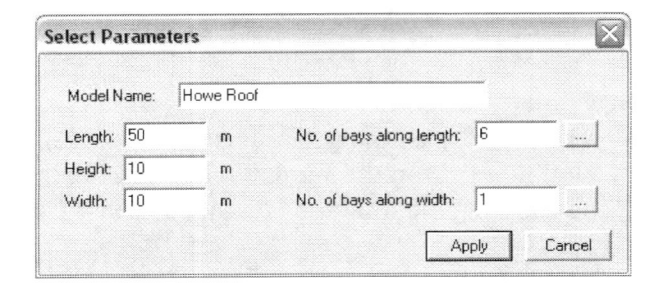

- As you can see from the dialog box, you can change the following parameters:
  - Total Length (in X direction).
  - Total Height (in Y direction).
  - Total Width (in Z direction), for 3D trusses only, if you want 2D truss set it to zero.
  - Number of bays in along length. This parameter will decide the shape of the truss.
  - Number of bays along width, set it to zero if you want 2D truss.
- The missing parameter is to control the number of bays in the Height (Y direction). This is not available because there are no spans in the Y direction.
- The rest of the procedure is the same as in the Frame Models.

Reference Point

- In previous sections we discussed how to create geometry in **Structure Wizard** and paste it in STAAD.*Pro* window but only if there is no structure. It is time to show how we can paste a geometry coming from Structure Wizard to an existing structure in the STAAD.*Pro* window. Do the following:
    - Create geometry in Structure Wizard.
    - Select **Edit/Add Paste Model** in STAAD.*Pro*, or click **Transfer Model** button from toolbar.
    - Confirm the transforming by clicking **Yes**. The dialog box shown below will be displayed.

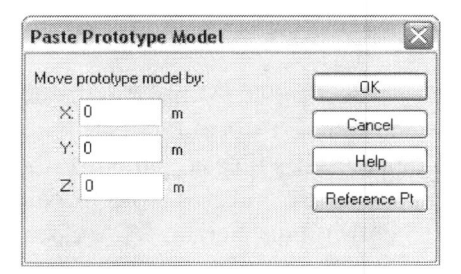

- You can input the XYZ coordinate right away, or (preferably) click on the **Reference Pt** (Pt means Point) button.
- The following screen will appear, asking you to specify the Node to handle created geometry from. Select one of the Nodes, and click **OK**.

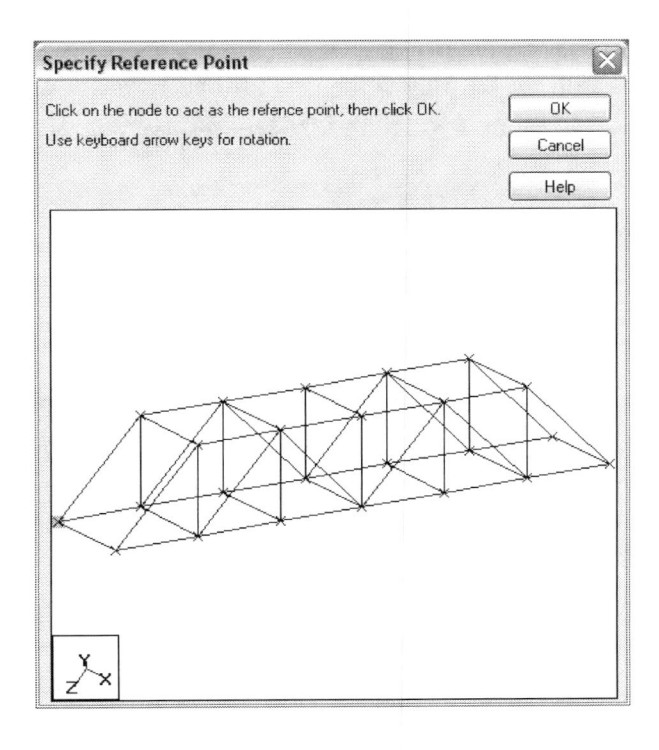

- The shape of the pointer will change to this shape.

- Click on the desired node at the structure in STAAD.*Pro* window. STAAD.*Pro* will return back to the old dialog box with the filtered coordinate of the needed point, as the dialog box shown below.

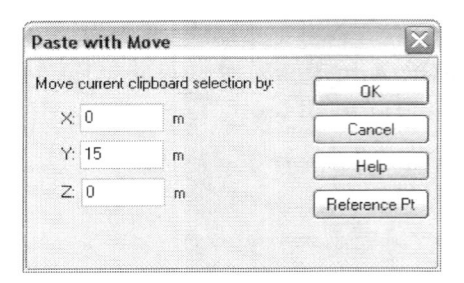

- Click **OK** to accept the results. Accordingly STAAD.*Pro* will display a message to inform the user that **Duplicate nodes ignored**, as shown below. This message means, those two nodes (one from the original structure and one from the created geometry) coincided in the same coordinate; hence, STAAD will ignore what is coming from the created geometry. Click **OK**.

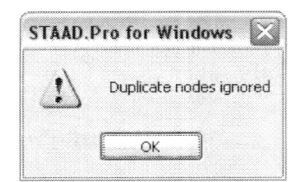

- The same issue applies to the beams; a new message will appear telling, **Duplicate beams ignored**. As shown in the dialog box shown below. Click **OK**.

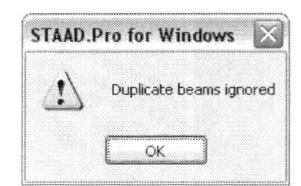

- Finally the geometry is pasted in the right place.

Using Structure Wizard to Create Truss & Using Reference Point

Exercise 3

1. Continue with the previous file.
2. Select **Geometry/Run Structure Wizard**.
3. Select Truss Models.
4. Double-click on Howe Roof icon.
5. Set the following parameters:
   a. Length = 9 m, cut to 4 bays as follows: 2+2.5+2.5+2.
   b. Height = 3 m.
   c. Width = 16 m, cut to 3 bays as follows: 5+6+5.
6. Select **Edit/Paste Model** in STAAD.*Pro*, or click Transfer Model icon from toolbar.
7. Confirm by clicking **Yes**.
8. Click **Reference Pt**. Make sure that the Reference point is on the far lower left. Click **OK** to accept it.
9. Click on the far upper left node of the frame. Confirm by clicking **OK**. Accept all the other messages.
10. The final structure should look like the structure in the next page.

The Final Structure

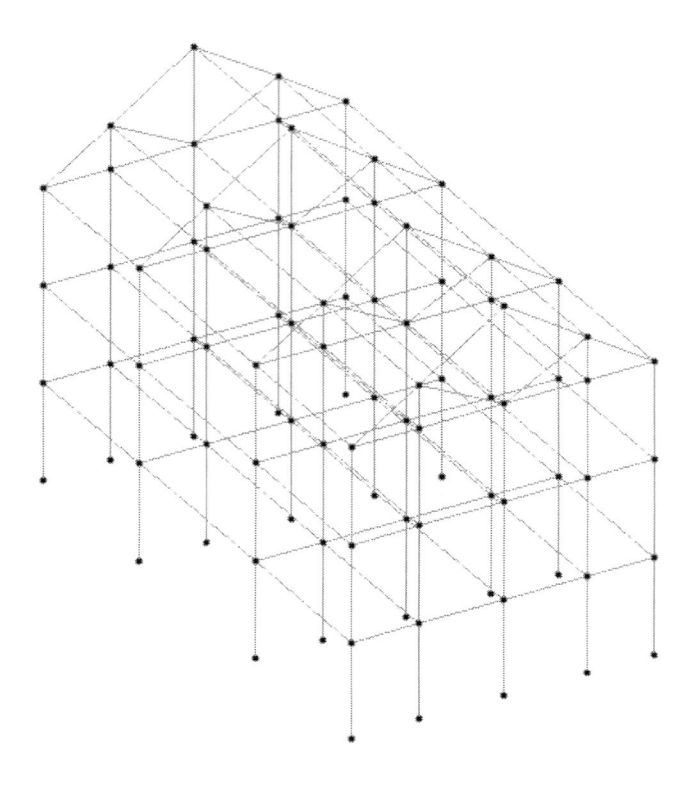

Surface/Plate Models / Quad Plate

- To create 3-noded, and 4-noded plates in any of three planes XY, XZ, and YZ. From Model pop-up list select **Surface/Plate Models**. Double-click on the **Quad Plate** icon; the following dialog box will be displayed:

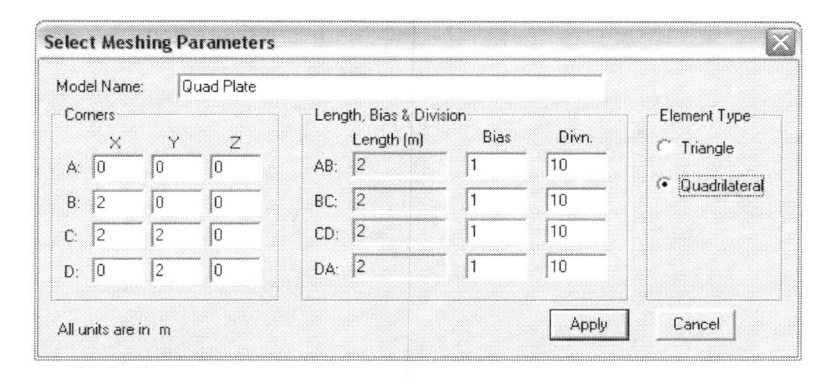

- From the **Element Type** (upper right portion of the dialog box) specify if you want **Triangle** shape (3-noded) or **Quadrilateral** shape (4-noded).
- You have 4 corners to specify A, B, C, and D, which they will be the corner of the desired plate. The XYZ here doesn't mean the real XYZ of the space, but rather XYZ of the Structure Wizard. The use of the XYZ is a very good way to tell Structure Wizard in which plane you will create your plate. As an example for the last note, check the following 4 corners:

    A = 0,0,0
    B = 6,0,0
    C = 6,0,6
    D = 0,0,6

- The result will be shown as the shape below:

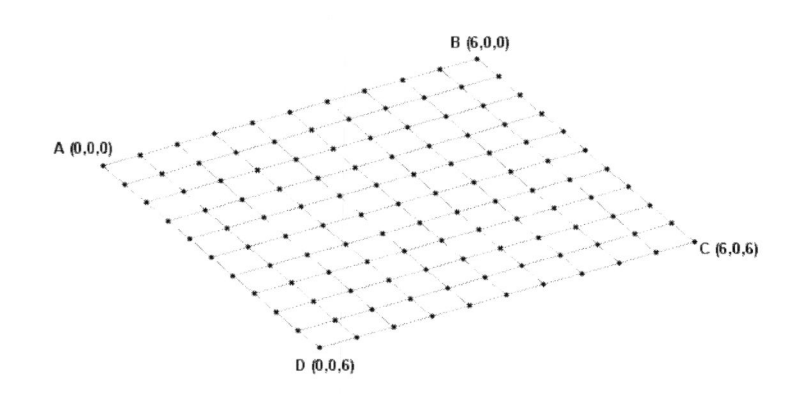

- As you can see the Y coordinate is always 0, hence the plate is in the X-Z plane, this is a good geometry for slab.
- As another example, check the following points:
  - A = 0,0,0
  - B = 0,4,0
  - C = 0,4,5
  - D = 0,0,5
-
- The result will be shown as the shape below:

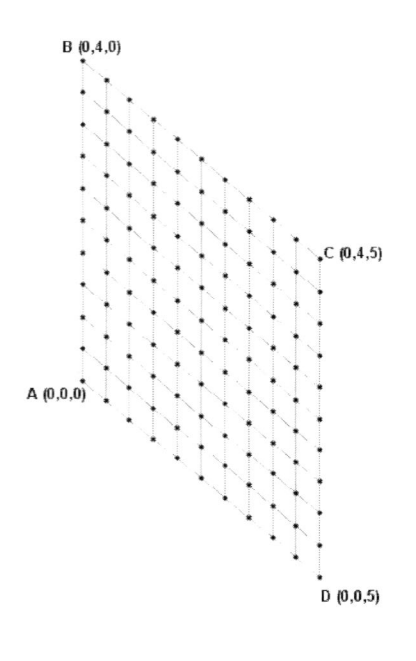

- Here X coordinate is 0; therefore the plane is Y-Z, a good setup for a wall.

- While giving the coordinates of the 3 or 4 nodes, you must be consistent, either rotate Clock Wise (CW), or Counter Clock Wise (CCW).
- STAAD.*Pro* will automatically calculate the length of each side.
- In the **Bias** and **Division** parts, specify the number of divisions each side of the plate will be divided to. By default Bias = 1, means the divisions are equally spaced. Dividing a plate means we will get more than one plate (one plate here means one entity). Example would be if you have a plate 6 X 6 m plate divided by 6 divisions from each side, therefore the total number of smaller plates will be 36 plates each is 1 m X 1 m.
- Click **Apply**. Then paste the plate on the structure existed in STAAD.*Pro* window using Reference Point as we learned in the previous section.

Note
- When you paste two 1X1 m plates on a 2 m beam, the plate will cut the beam into two beams each one is 1 m length, by creating a node in the middle of the beam. See the illustration below.

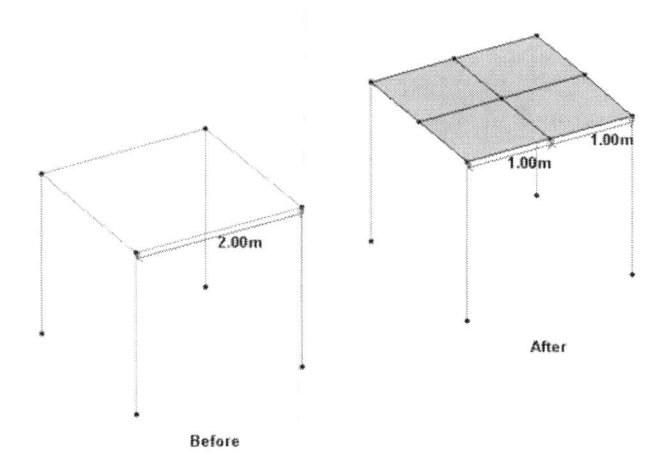

Using Structure Wizard to Plates

Exercise 4

1. Continue working in the file of last exercise.
2. Create a Quadrilateral plate with the following information:
   a. A = 0,0,0
   b. B = 2,0,0
   c. C = 2,0,5
   d. D = 0,0,5
   e. AB Division = 2
   f. BC Division = 5
   g. CD Division = 2
   h. DA Division = 5
   i. Bias is always = 1
3. Paste it in a point to look like figure in the next page. (Hint the point on the structure should be 0,3,0)

The Final Structure

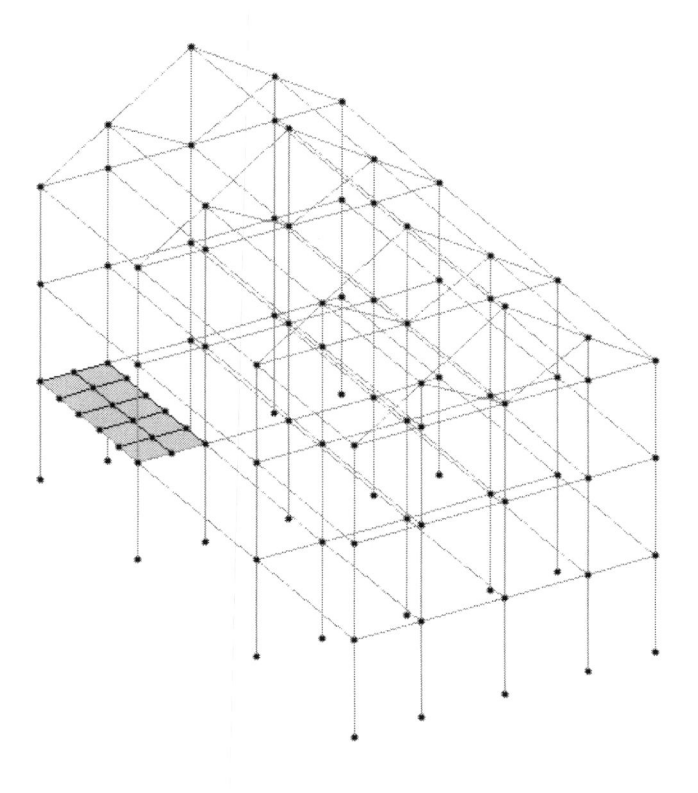

Things you can do in Structure Wizard

- While you are in the Structure Wizard, several viewing functions are available to help you visualize your *model-to-be-transferred*. Some of these functions are available also in the STAAD.*Pro* window, maybe with some additions. These functions are available after you create a model and before you transfer it. They are:

- From the toolbars click on **Toggle Axes View,** or select **View/View Axes**. This function can be switched ON, or OFF, and its purpose is to show, or hide the XYZ icon (X=red, Y=green, Z=blue) representing the 3 Axes in Structure Wizard.

- From the toolbars click on **Toggle Perspective View**, or select **View/Perspective**. There are two possible views in Structure Wizard, either Isometric, or Perspective (default), if this toggle is ON, it will show Perspective. Check the illustration below:

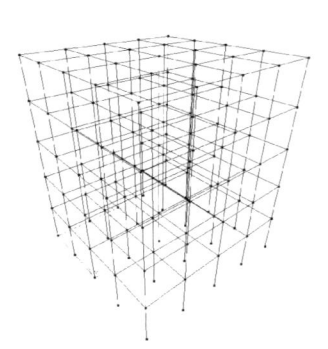

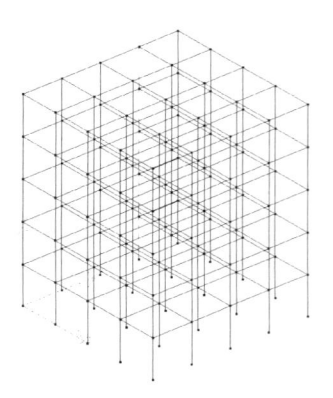

- From the toolbars click on **Toggle View Mode** or select **View/Wireframe View**. This is very useful in the Surface/Plate Models. You can select between Wireframe view and Solid Fill view. Check the two images below:

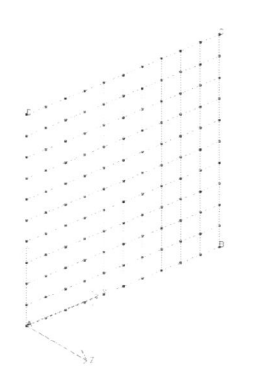

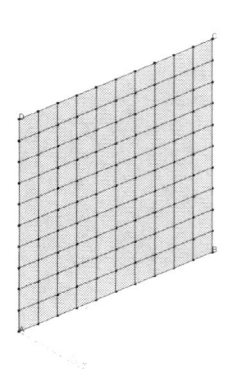

2-28

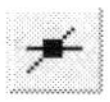

Module 2: Geometry

- From the toolbars click **Toggle Node Markers**, or select **View/View Nodes**. This toggle is to display the node markers or not. See the illustration below:

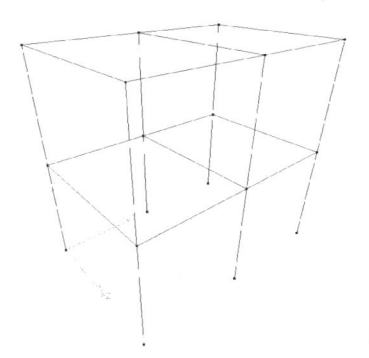

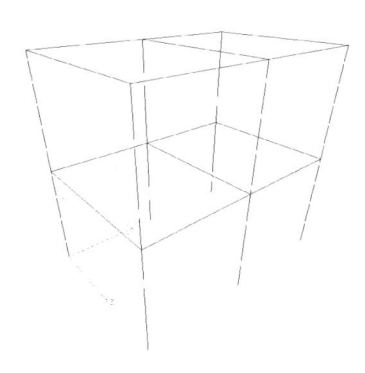

- From the toolbars, click **XY View (Elevation)**, or select **View/ Elevation (XY) View**. This will show 2D view, to see XY plane of the structure. See the illustration below (this is a structure of X=6m, Y=15m, Z=12m and shown as Perspective, and will be shown in all illustration below):

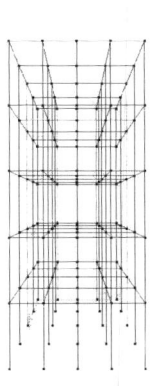

- From the toolbars, click **YZ View (Side)**, or select **View/Side (YZ) View**. This view will show 2D view, to see only the YZ plane of the structure. Check the illustration below:

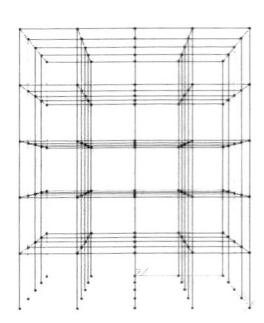

- From the toolbars, click **XZ View (Top)**, or select **View/Top (XZ) View**. This view will show 2D view, to see only the XZ plane of the structure. Check the illustration below:

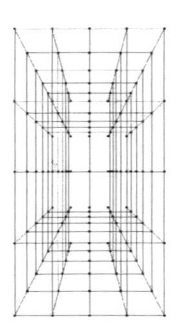

- From the toolbars, click **Isometric View**, or select **View/Isometric View**. This view will show 3D view. Check the following illustration:

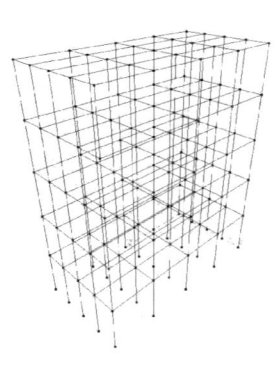

Using Structure Wizard Viewing Commands

Exercise 5

1. Continue working in the file of last exercise.
2. Start Structure Wizard and create any Frame you want.
3. Using the four toggles you learned, try to use them and see the effect of each one of them.
4. Using the four viewing points, try to use them and see the effect of each one of them.

Method 2: Drafting the Geometry using Snap/Grid

- Using the **Snap/Grid** utility provided by STAAD.*Pro*, the user is capable of drafting the structure needed.
- Before doing any thing we have to understand what are the steps to prepare STAAD.*Pro* window to allow the user start drafting. These steps are:
- Initiate the **Snap/Grid**. There are three ways to do that:
  - Select from the Page Control **Geometry** tab.
  - Select from the menus **Geometry/Snap Grid Node** then **Beam**, **Plate**, or **Solid**.
  - From the **Geometry** toolbar select one of two icons available, **Snap Node/Beam**, or **Snap Node/Plate**.
- Either way a dialog box will appear in the Data Area, like below:

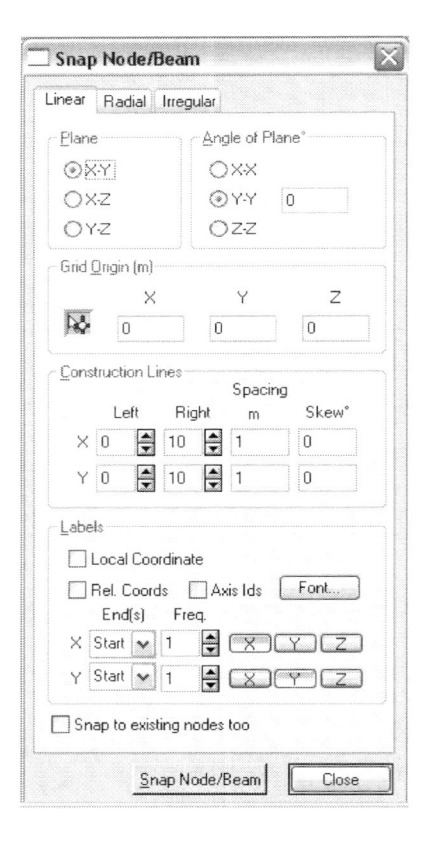

2-32

- Also you will see in the STAAD.*Pro* window, a grid in XY plane like the following image.

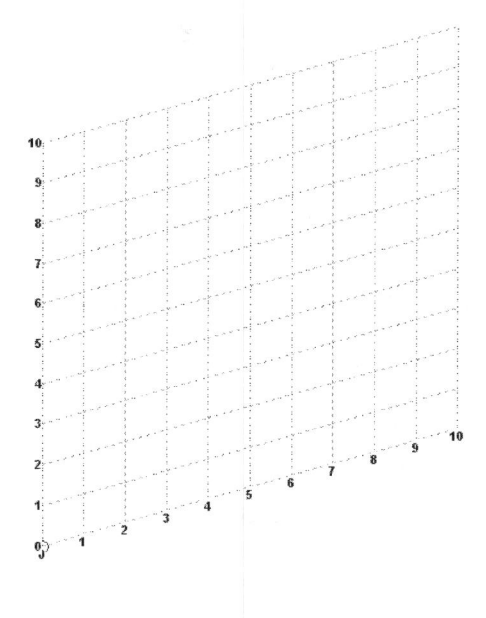

- Decide in which plane you want to work, XY plane, XZ, or YZ.
- Specify Angle of Plane (Leave it 0 for now).
- Specify Origin (preferable to leave it at current 0,0,0).
- Specify the Construction Lines, take care of the following points:
  - If you want the Origin to be 0,0,0 make sure that the **Left** value for X, and Y is always 0, this will make sure that the lower left corner of the Grid is always 0,0,0.

  - In the part labeled **Right** (for both X, and Y), input the *number of Grid segments* in that axis.

  - Under **Spacing**, there are two fields to be filled, **m**, and **Skew**. In the **m** part, input the length of segments of the Grid. As an example to the above two points: assume you input in the part labeled **Right** in X direction 10, and in the **m** part you input 2, the total length is 10X2=20 m.

Note	■ Note the following in the STAAD.*Pro* window: • A moving black bold cross (let's call it Controlling Point) following the steps of the cursor. • The coordinates of that cross appear in the right portion of the status bar like below. X: 2.000 Y: 0.000 Z: 0.000 • The circle, which appears at the lower left corner of the Grid, which represent the origin.
Adding Beams	■ Make sure that **Snap Node/Beam** is on. ■ To start drafting **Beams**, go to the start Node coordinate and click, a Node will be inserted there. Go to the next Node, and click, a second new Node will be added and accordingly a new Beam will be created. Keep on doing this until you are done, then click **Close** in dialog box.
Note	■ Once you start clicking Nodes, the Controlling Point will strict you to start your next Beam from the last Node reached. In order to avoid this, hold **Ctrl** key at the keyboard, and click on the Node Coordinate desired other than the last Node and you can start your next Beam from that Node.
Adding Plates	■ Make sure that **Snap Node/Plate** is on. ■ Go to the start desired coordinate and click, a Node will be added there, repeat this process for four points, a new Plate will be added. When you are done click **Close**.
Note	■ This way will always draw 4-Noded Plates. ■ Once you finish the first plate, the Controlling Point will strict you to start your next Plate from the last Node reached. To avoid this, hold **Ctrl** key at the keyboard, and click on the coordinate desired other than the last Node and you can start your next Plate from that Node.

2-34

Fill Plates ■ To view your plates better, make sure to do the following:
- In the STAAD.*Pro* window, right-click anywhere, shortcut menu will appear, select from it **Structure Diagrams**. The following dialog box will appear:

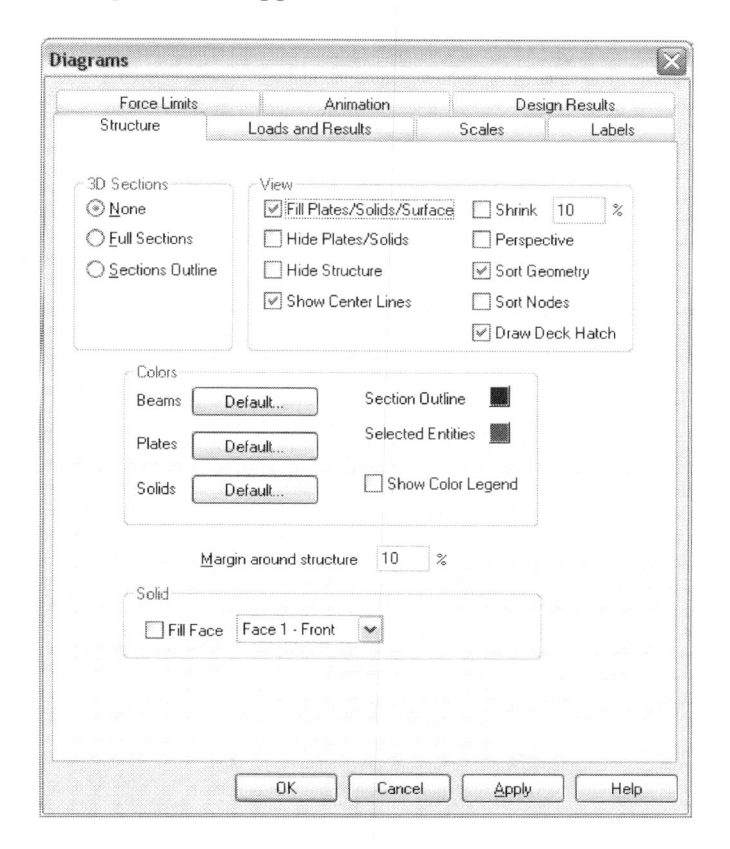

- Under **View**, click **Fill Plates/Solids/Surfaces** ON, click **Sort Geometry** ON, click **Sort Nodes** ON.

Using Snap/Grid

Exercise 6

1. Start a New Space Frame file.
2. Select the **Geometry** tab at the **Page Control**.
3. Make sure that you are using the X-Y Plane.
4. In the **Construction Lines** part, input the following data:
    a. For X, Left=0, Right=3, m=4.
    b. For Y, Left=0, Right=2, m=3.
5. Make sure that **Snap Node/Beam** is ON.
6. Click the following coordinates (use coordinates displayed in the status bar to help you):
    a. 0,0,0
    b. 0,3,0
    c. 12,3,0
    d. 12,0,0
    e. Hold Ctrl key and click 4,3,0
    f. 4,0,0
    g. Hold Ctrl key and click 8,3,0
    h. 8,0,0

7. By now your model should look like the following image:

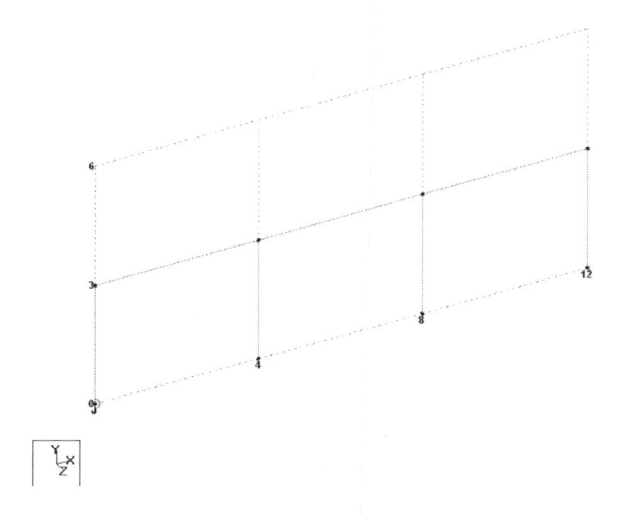

8. Change the **Plane** to X-Z plane.
9. Change the **Origin** to 0,3,0; the Grid should be elevated to top of the frame.
10. In the **Construction Lines** part, keep X values as is. Change the Z values to be Left=0, Right=1, m=4.
11. Click the following coordinates:

    a. 0,3,0

    b. 0,3,4

    c. 12,3,4

    d. 12,3,0

    e. Ctrl + 4,3,0

    f. 4,3,4

    g. Ctrl + 8,3,0

    h. 8,3,4

12. Your model should look like this:

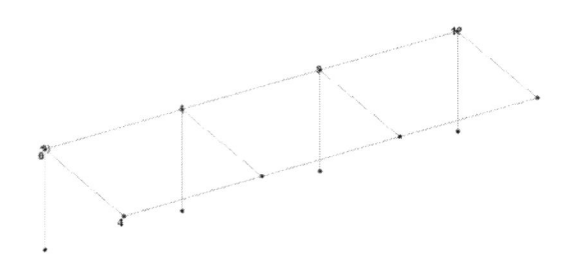

13. Using the same methods discussed in this exercise try to create additional members to make the structure look like this:

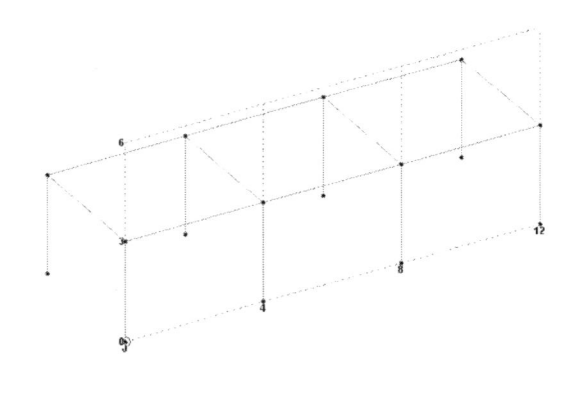

14. Close Snap Node/Beam.
15. Using the **Geometry** toolbar, click **Snap Node/Plate/Quad** to draft **Plates** instead of **Beams**.
16. Change the **Plane** to X-Z plane.
17. Change the **Origin** to 0,3,0.

18. In the **Construction Lines** part:

    a. X values are Left=0, Right=3, m=4.

    b. Z values are Left=0, Right=1, m=4.

19. Click the following coordinates in the same sequence (plates should be drafted either CW, or CCW, you can't use the zigzag method).

    a. 0,3,0

    b. 0,3,4

    c. 4,3,4

    d. 4,3,0

    e. 8,3,0

    f. 8,3,4

    g. 4,3,4

    h. Ctrl + 8,3,0

    i. 12,3,0

    j. 12,3,4

    k. 8,3,4

20. Three green plates are drafted now as shown below, Click Close.

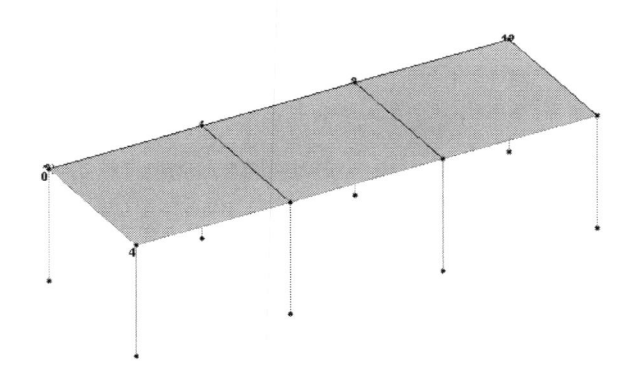

Before we go on with the rest of the methods to create geometry, we have to discuss two important functions, which will help us accomplish the rest of the methods swiftly. These two functions are: **Viewing** your geometry and **Selecting** Nodes, Beams, and Plates.

Viewing

- In previous part of this tutorial we went through the four viewing functions in Structure Wizard, these four and three more are available in STAAD.*Pro*.
    - View from +Z (It is Elevation in Structure Wizard). You can consider it the Front view.
    - View from –Z, is the Back view.
    - View from –X, is the Left view.
    - View from +X, (It is Side in Structure Wizard). You can consider it the Right view.
    - View from +Y, (It is Top in Structure Wizard).
    - View from –Y, is the Bottom view.
    - Isometric, is the isometric view.
- We have 6-rotation function, which capable of rotating the geometry around a specific axis, these are:
    - Rotate Up & Rotate Down (Rotating around X in both directions).
    - Rotate Left & Rotate Right (Rotating around Y in both directions).
    - Spin Left & Spin Right (Rotating around Z in both directions).

Note
- You can use the arrows in your keyboard also. Use:
    - *Right* arrow and *Left* arrow to rotate around Y-axis.
    - *Up* arrow, and *Down* arrow to rotate around X-axis.

Selecting

- You need to select either a Node, Beam, or Plate in order to perform a command on them. As a first step of selecting any thing in STAAD.*Pro* choose the right cursor. 
  - To select a Node choose the Nodes Cursor. 
  - To select a Beam choose a Beams Cursor. 
  - To select a Plate choose a Plates Cursor.
- After you choose the right cursor, you have four ways; they are:

Single Selection
  - Click on the desired Node, Beam, or Plate, it will be highlighted by turning into red. Check the figure below:

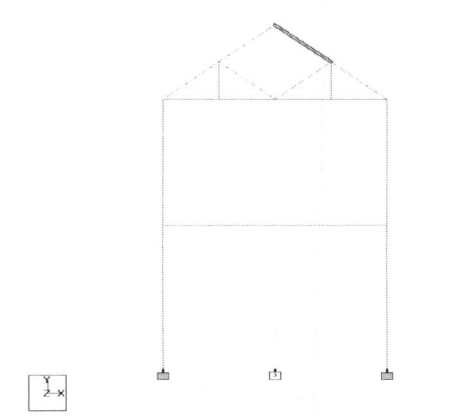

  - From the Data Area, click on the number of the Node, Beam, or Plate, it will be highlighted. Check the figure below:

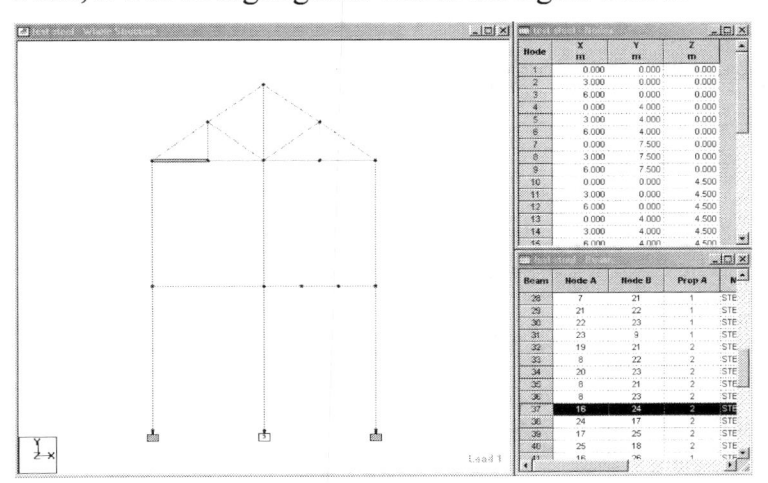

Multiple Selection
- Select the first Node, Beam, or Plate, then hold down the **Ctrl** key at the keyboard, and click other Nodes, Beams, and Plates. Check the figure below.

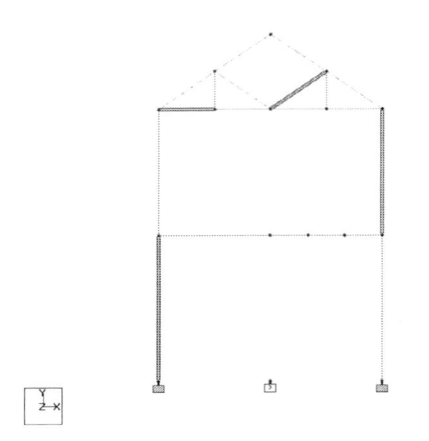

- From the Data Area, click on the number of the Node, Beam, or Plate, it will be highlighted. Then hold the **Ctrl** key at the keyboard, and click on other numbers; it will be highlighted as well. Check the figure below.

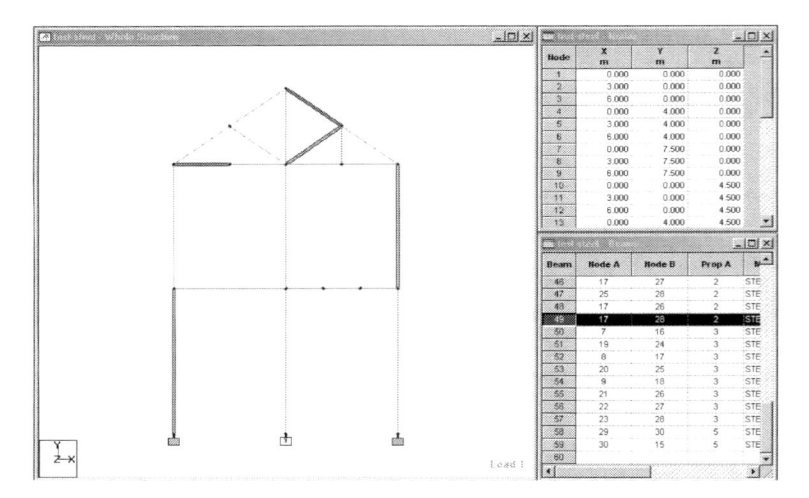

- Make a Window around the needed Nodes; Beams, or Plates, by clicking in an empty place of the STAAD.*Pro* window, and holding down the left button, moving to the other corner and releasing the button, what ever inside the Window will be selected automatically. Check the figure below:

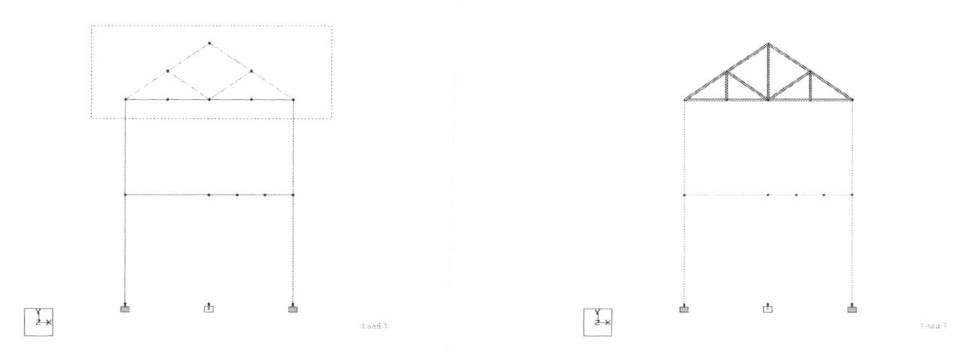

Note ■ As for Beams, the *Mid Point* of the Beam is the important part that should be included in the Window. Check the illustration.

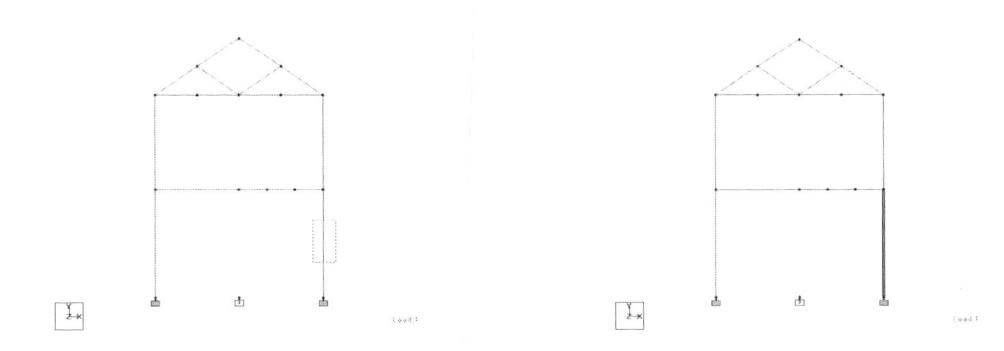

Ctrl+A ■ To select all Nodes, Beams, or Plates, first select the proper cursor, and the press **Ctrl+A**, every thing will be selected accordingly.

Unselect ■ To unselect any selected Nodes, Beams, or Plates, simply click on an empty space.

Using Selecting while viewing 3D Geometry

- Using both Viewing commands and Selecting methods leads to effectively select multiple Nodes, Beams, or Plates, in 3D Geometry. Looking at a 3D model from different viewing points will enable the user to select Nodes, Beams, and Plates in the plane shown and any things behind it.

Example
- Check below figure, which represents a 3D geometry.

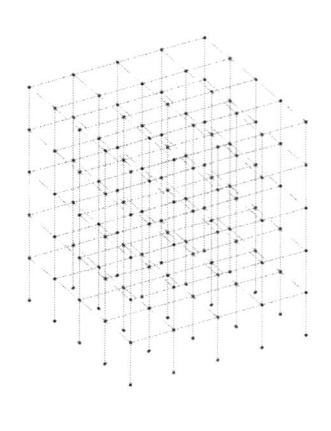

- Click on **View From +Z icon**, check the result in below figure:

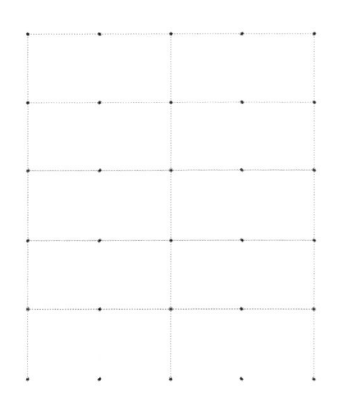

- Now click on one of the Beam as shown:

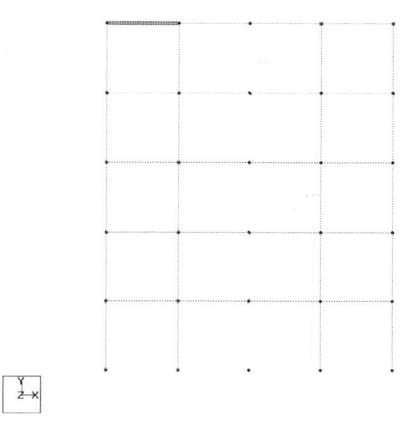

- Click the Isometric view, to see the result:

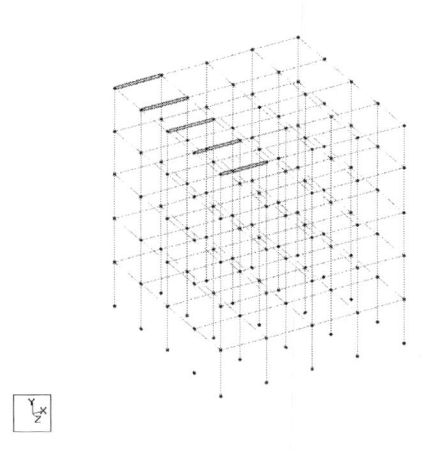

Note ■ To speed up the selection method of 3D geometry, use Window, which will enable the user to select multiple Nodes, Beams, and Plates, in the 2D View.

Viewing & Selecting

Exercise 7

1. Create a new Space Frame file
2. From Structure Wizard, create the following structure:
   a. Length = 12 m   # of Bays = 4
   b. Height = 15 m   # of Bays = 5
   c. Width = 12 m   # of Bays = 4
3. Change the cursor to Nodes cursor.
4. Using the **View From +Z**, and using Window select the upper Nodes.
5. Click anywhere to unselect.
6. Change the cursor to Beam Cursor.
7. Using the **View From +Y**, and using cursor and Ctrl key select all the horizontal Beams.
8. Click anywhere to unselect.
9. Press Ctrl+A.
10. Change the cursor to Nodes cursor, then press Ctrl+A. What is the difference?

Method 3: Using Copy/Cut with Paste

- The first two methods are meant to creating geometry from the scratch, but this method is to create a copy of an existing geometry. You can copy Nodes, Beams, and Plates. Of course when you are copying Beams, and Plates STAAD.*Pro* will copy the associated Nodes as well.

Steps
- Select the desired objects to copy (Nodes, Beams, or Plates) making sure that you are using the right cursor.
- From the menus choose **Edit/Copy**, or press **Ctrl+C** (you can use also the **Edit/Cut** or **Ctrl+X**, but this will be considered as moving).
- From the menus choose **Edit/Paste**, or press **Ctrl+V** (if the selected objects are Nodes, it will show **Paste Nodes**, and if Beams, it will show **Paste Beams**, and so on). The following dialog box will be displayed to enable the user to paste the selected objects in the right place.

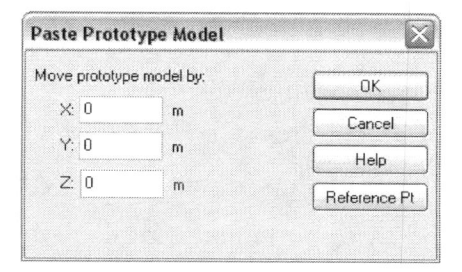

- This dialog box is the same as the one we dealt with in pasting a geometry coming from Structure Wizard, hence all things discussed there is applicable here.

Creating Geometry with Copy/Cut and Paste

Exercise 8

1. Create a new Space Frame file.
2. From Structure Wizard, create the following structure:

    a. Length = 6 m   # of Bays = 1

    b. Height = 3 m   # of Bays = 1

    c. Width = 4 m   # of Bays = 1

3. Press **Ctrl+A** to select all Beams.
4. Choose **Edit/Copy** or press **Ctrl+C** to copy Beams.
5. Choose **Edit/Paste Beams**, or press **Ctrl+V** to paste Beams.
6. Type in the following coordinates X=0, Y=3, Z=0 (also you can use Reference Pt)
7. The final shape should look like the figure below:

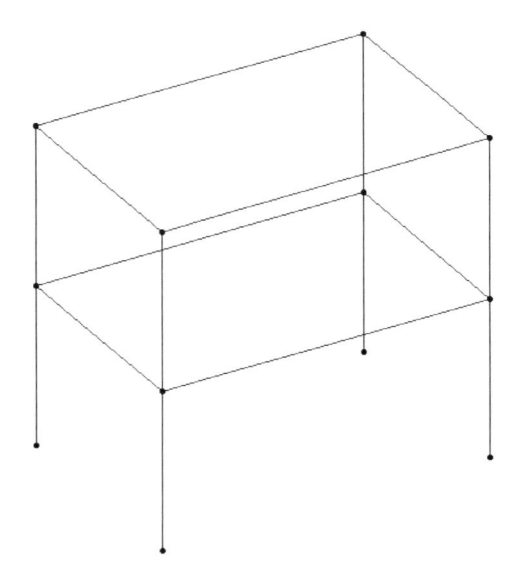

Method 4: Using Spreadsheet (Excel) Copy and Paste

- To use this method you have to have a good knowledge of how to use Excel, and how to write formulas. Our main purpose here is to generate non-conventional geometries, which involve mathematical equations. Excel will produce points, and we will copy them using OLE to STAAD.*Pro* to generate Nodes.

Steps
- Start Excel program (or any spreadsheet software).
- Type in any mathematical formula and generate X, Y, Z points (as much you increase the number of points, you may get better geometries specially if there are curves in the structure).
- Select the columns representing X, Y, Z (without any headings)
- From Excel menus choose **Edit/Copy**.
- Go to STAAD.*Pro*.
- At the Geometry Page Control, and while you are inside the Nodes table, select the first node number by clicking on the Node number.

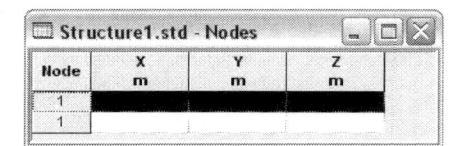

- Choose **Edit/Paste** or right-click and choose Paste. The following dialog box will appear:

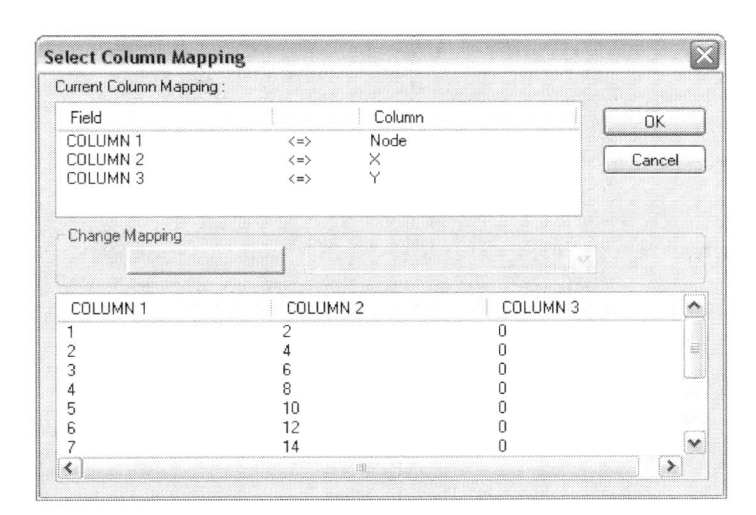

- Select to map the first column as X, the second column as Y, and the third column as Z, then click OK, the new Nodes will be added accordingly.

Add Beams
- The previous function will help us add Nodes only. We need to use the **Add Beams** function, to link the Nodes. 
- From the Geometry Toolbar, click the Add Beams tool, or from Menu select Geometry/Add Beam/Add Beam from Point to Point.
- The mouse shape will change to this shape. Click on the first Node, a rubber band will appear waiting for the second Node, click the second Node, and you will have a new Beam added. Repeat this process up until you finish the whole Nodes.

Note
- Add Beams can help the user to make the bracing for Frames.

Add 3-Noded Plates
- Use **Add 3-Noded Plates**, to link Nodes with triangular plate. 
- From the **Geometry** Toolbar, click the **Add 3-Noded Plates** tool, or from Menu select **Geometry/Add Plate/Triangle**. 
- The mouse shape will change to this shape. Click on the first Node, second Node, and third Node you will have a new 3-Noded Plate.

Add 4-Noded Plates
- Use **Add 4-Noded Plates**, to link Nodes with quadratic plate. 
- From the **Geometry** toolbar, click the **Add 4-Noded Plates** tool, or from Menu select **Geometry/Add Plate/Quad**. 
- The mouse shape will change to this shape. Click on the first Node, second Node, third Node, and fourth Node.

Using Labels
- In the STAAD.*Pro* Window, right-click a shortcut menu will appear, select from it **Labels**, a large dialog box will appear, turn on **Node Numbers**, **Node Points**, **Beam Numbers**, and **Plate Numbers**.

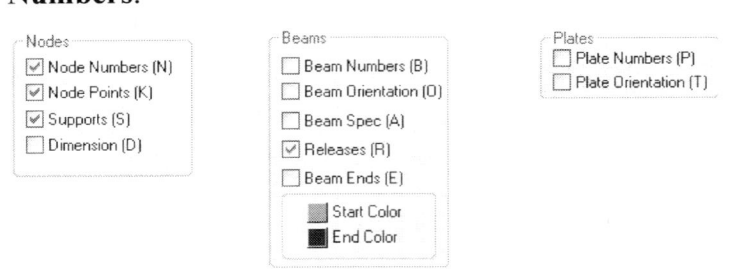

Nodes:
- ☑ Node Numbers (N)
- ☑ Node Points (K)
- ☑ Supports (S)
- ☐ Dimension (D)

Beams:
- ☐ Beam Numbers (B)
- ☐ Beam Orientation (O)
- ☐ Beam Spec (A)
- ☑ Releases (R)
- ☐ Beam Ends (E)
- ■ Start Color
- ■ End Color

Plates:
- ☐ Plate Numbers (P)
- ☐ Plate Orientation (T)

Note
- Turning on numbers may lead to make the picture of the structure cluttered, so be careful to *pick-and-choose*.

Creating Geometry Using Excel Copy & Paste

Exercise 9

1. Start Excel, in a new sheet, make the following table:

X	Y	Z
0.00	25.00	0.00
0.50	24.75	0.00
1.00	24.00	0.00
1.50	22.75	0.00
2.00	21.00	0.00
2.50	18.75	0.00
3.00	16.00	0.00
3.50	12.75	0.00
4.00	9.00	0.00
4.50	4.75	0.00
5.00	0.00	0.00

Y formula is $25-X^2$; where X is the cell address containing X values. Don't forget to use all the copying functions of Excel.

2. Start STAAD.*Pro*, and create a new Space file.
3. Copy the table you made in Excel to the Node table in the STAAD file.
4. Map the first column to be X, 2nd to be Y, and 3rd to be Z.
5. From **Labels**, turn on the **Node Points**.
6. Using the **Add Beams**, add the necessary beams.
7. The resulting geometry should look like:

The Final Structure

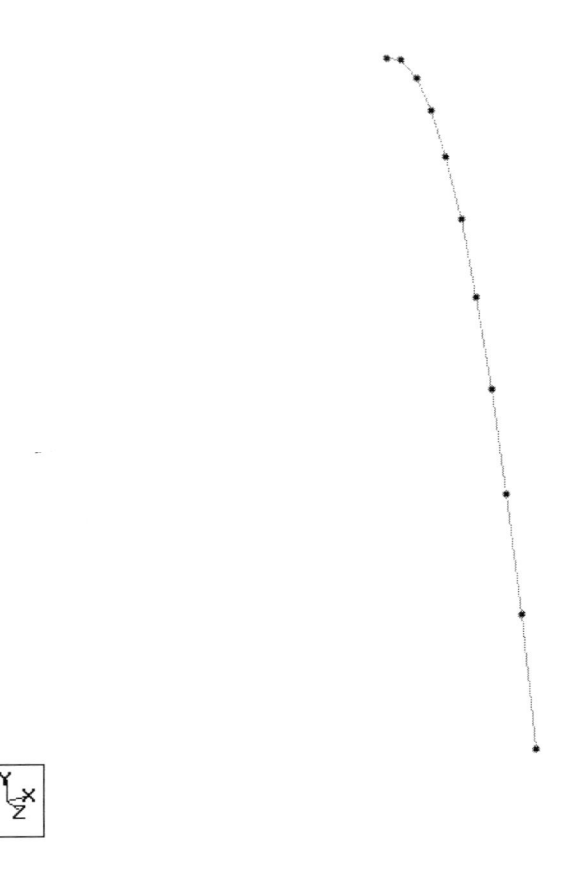

Method 5: Using DXF importing file function

- To use this method you have to be professional CAD (or specifically AutoCAD) user, in order to produce 2D, or 3D geometries.

Steps
- Start AutoCAD (or any CAD that can produce DXF file).
- Draw your structure 2D or 3D.
- **Save As** DXF file.
- Start STAAD.*Pro*.
- In order to read DXF file, you have two methods:
  - From the Structure Wizard.
  - From **File/Import**.

From Structure Wizard
- Start Structure Wizard.
- From the Model Type select **Import CAD Models**.
- Double-click on the **Scan DXF** icon; a dialog box will appear so you can select the DXF file name:

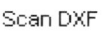

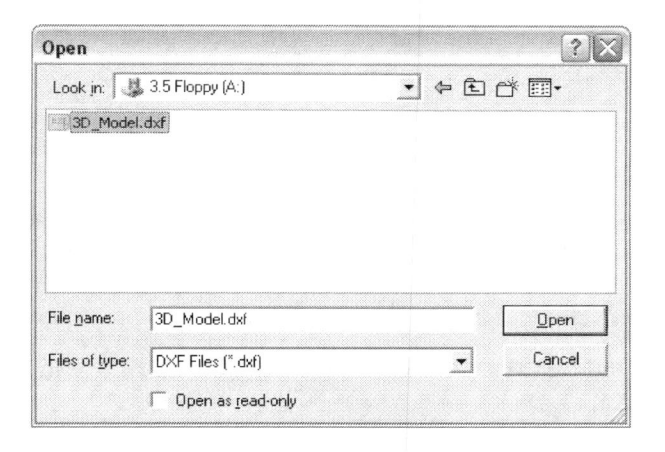

- Select the desired file, and click **Open**.

Note
- The DXF will be scanned, but *without* rotating it correctly.

From File/Import
- From File menu, select **File/Import**.
- STAAD.*Pro* asks now about the source of the file to be imported:

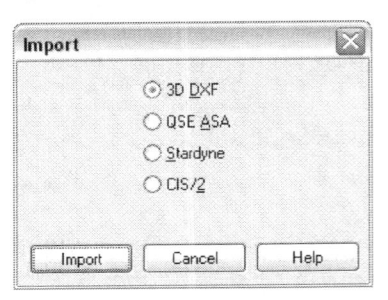

- Select **3D DXF**, and click **Import**. STAAD.*Pro* will ask for the location of the DXF file:

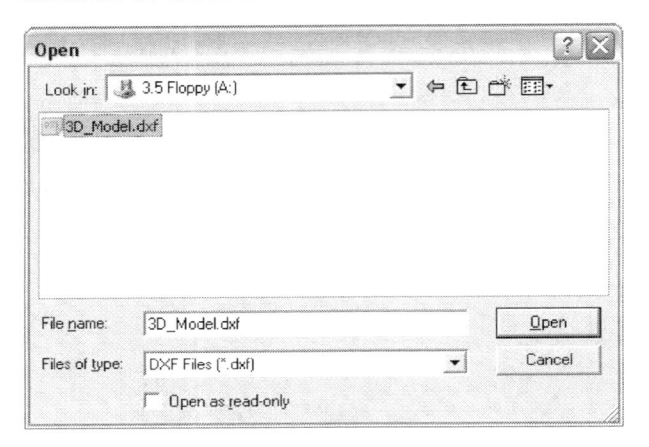

- Select the file, and click **Open**. The following dialog box will appear:

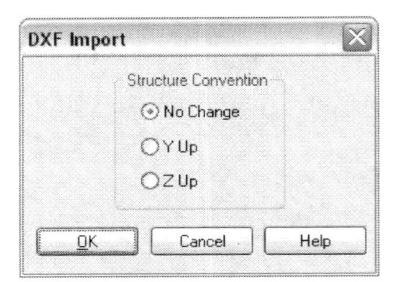

2-54

- Now select one of the three choices, and click **OK**:
  - **No Change**; the XYZ orientation of STAAD matches the XYZ in AutoCAD.
  - **Y Up**; you are telling STAAD to consider Y is up in STAAD, and hence to convert Y in AutoCAD accordingly. (This is the right choice in almost all of the cases)
  - **Z Up**; you are telling STAAD to consider Z is up in STAAD, and hence to convert Z in AutoCAD accordingly.
- The following of the dialog box will appear:

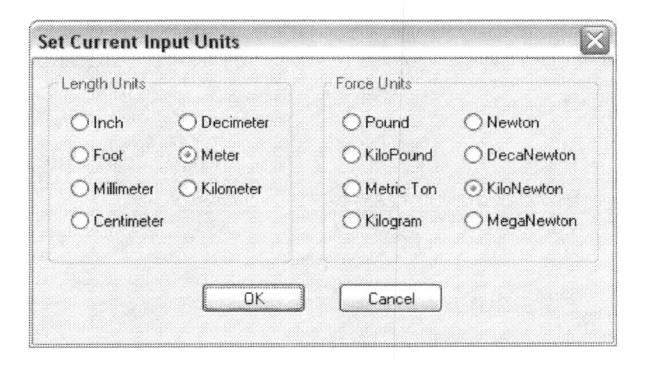

[Set Current Input Units dialog box with Length Units (Inch, Decimeter, Foot, Meter (selected), Millimeter, Kilometer, Centimeter) and Force Units (Pound, Newton, KiloPound, DecaNewton, Metric Ton, KiloNewton (selected), Kilogram, MegaNewton), OK and Cancel buttons]

- Select the proper **Length Unit**, and the proper **Force Unit**, and click **OK**, the structure will be transferred.

Note
- In AutoCAD use always **Line**, in drafting Beams and Columns.
- STAAD will consider one Line; equal to one Beam or Column, hence, long line covering more than one Node will be considered as one object, accordingly cut your lines on the intersections.
- Use the latest AutoCAD version with the latest STAAD versions.

Creating Geometry Using DXF importing file function

1. Create any structure in AutoCAD, and save as DXF file.
2. Start STAAD.*Pro*, and create a new Space file.
3. Select **File/Import**, select **3D DXF**, and specify the file you created.
4. Select **Y Up**.
5. Select **Meter**, and **KiloNewton**, and click OK, the structure will be imported to STAAD.*Pro* window.
6. See how the DXF file turns to be a STD file.
7. Save and Close.

Creating Geometry (Concrete Structure)

Workshop 1-A

1. Create a new folder in **C:**, and name it **Concrete_Model**.
2. Start STAAD.*Pro*.
3. Create a New File using the following data:
   a. **File Name**: Small_Building
   b. **Location**: C:\Concrete_Model
   c. **Type of Structure**: Space
   d. **Length Unit**: Meter, **Force Unit**: KiloNewton
   e. Click **Next**, Select **Edit Job Info**, and click **Finish**.
4. In the Job Info, fill the following data:
   a. **Job**: Small Commercial Building.
   b. **Client**: ABC Investment Corp.
   c. **Job No.**: 112-8101965
   d. **Ref**: A-114
   e. Under the Engineer Name, put your initials, and leave the rest empty.

5. Using any method you know, create the following frame:

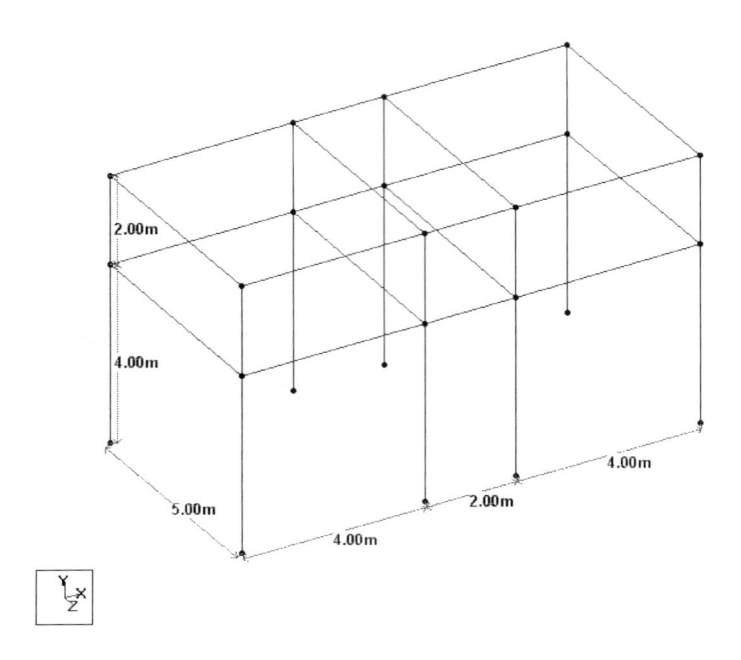

6. Using any method you know add the plates as shown: (Note each plate is 1 m x 1 m).

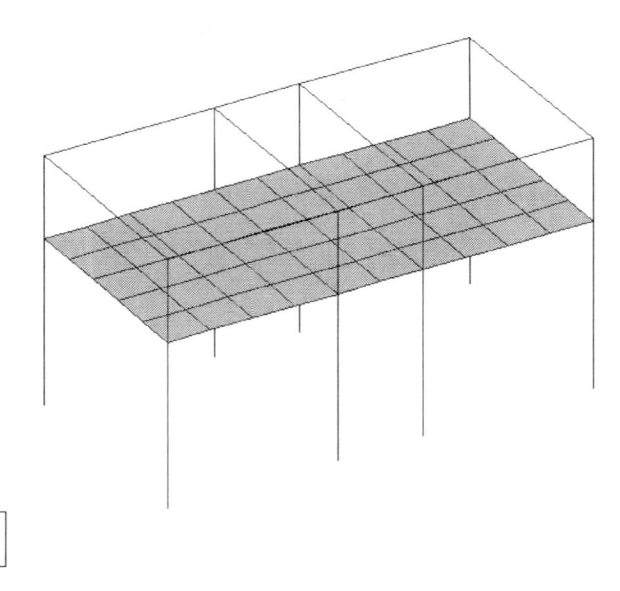

Module 2: Geometry

7. Copy the plates to the upper part, to get the following geometry:

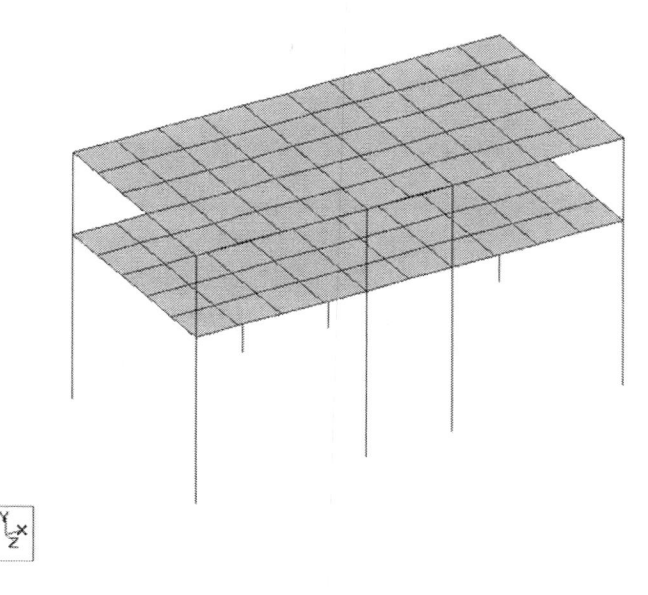

8. Save and close.

Creating Geometry (Steel Structure)

Workshop 1-B

1. Create a new folder in **C:**, and name it **Steel_Model**.
2. Start STAAD.*Pro*.
3. Create a New File using the following data:
    a. **File Name**: Small_Building
    b. **Location**: C:\Steel_Model
    c. **Type of Structure**: Space
    d. **Length Unit**: Meter, **Force Unit**: KiloNewton
    e. Click **Next**, Select **Edit Job Info**, and click **Finish**.
4. In the Job Info, fill the following data:
    a. **Job**: Small Commercial Building.
    b. **Client**: ABC Investment Corp.
    c. **Job No.**: 112-8101965
    d. **Ref**: A-114
    e. Under the Engineer Name, put your initials, and leave the rest empty.

5. Using any method you know, create the following frame:

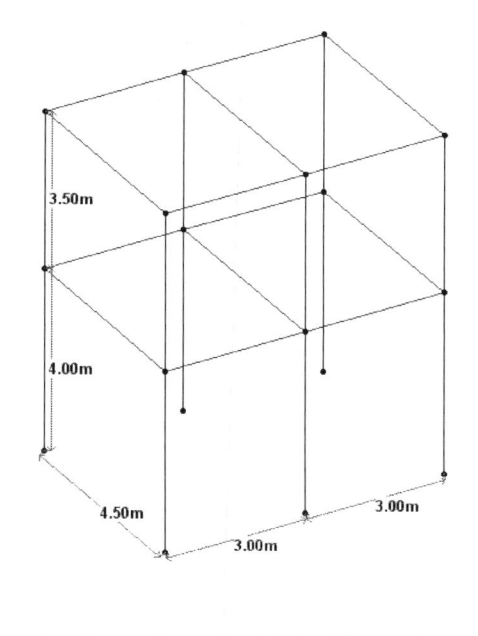

6. Using any method you know add the Howe Roof as shown:

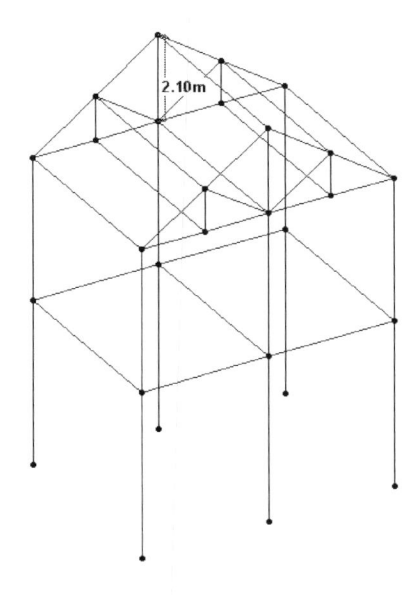

Notes:

Module Review

1. In the input file and under Joint Coordinate, one of the following statements is true:
    a. 1 0 0 0
    b. 1,0,0,0
    c. 1;0;0;0
    d. 1 0 0

2. In STAAD.*Pro*, Nodes are the geometry and not the Beams:
    a. True
    b. False

3. There are _____ ways to build up a geometry in STAAD.*Pro*

4. I can import AutoCAD DWG file to STAAD.*Pro*
    a. True
    b. False

5. What is Structure Wizard:
    a. Brings Excel sheet to STAAD.*Pro*
    b. Allows user to import AutoCAD DWG files
    c. A library of pre-defined structural shapes
    d. All of the above

6. With Labels you can switch on _____, _____, _____ features.

Module Review Answers

1. a
2. a
3. 5
4. b
5. c
6. Node Numbers, Beam Numbers, Plate Numbers, Node Points, etc.

Module 3:

Useful Functions to Complete the Geometry

This Module contains

- Translational Repeat
- Circular Repeat
- Mirror
- Rotate
- Move
- Insert Node
- Adding Beams (Connecting & Intersecting)
- Cut Section
- Renumber
- Miscellaneous Functions

Module 3: Useful Functions to Complete the Geometry

Introduction

- The five methods we discussed in Module 2 are used to create the basic geometry. However, cannot alone fulfill the creation of some complex requirements of structural engineer.
- In this module, we will discuss essential functions, which will enable the user to complete any unusual requirements in building up the geometry.
- User should select Node, Beam, or Plate before issuing any of the functions to be discussed herein.
- When you combine Module 2 & Module 3, you will know all the geometry function exists in STAAD.*Pro*.

Translational Repeat

- With this function we can duplicate Nodes, Beams, Plates in the direction of X, Y, or Z.
- Select the desired Nodes, Beams, or Plates to be duplicated.
- From **Generate** toolbar, select **Translational Repeat**, or from menus select **Geometry/Translational Repeat**, the following dialog box will appear:

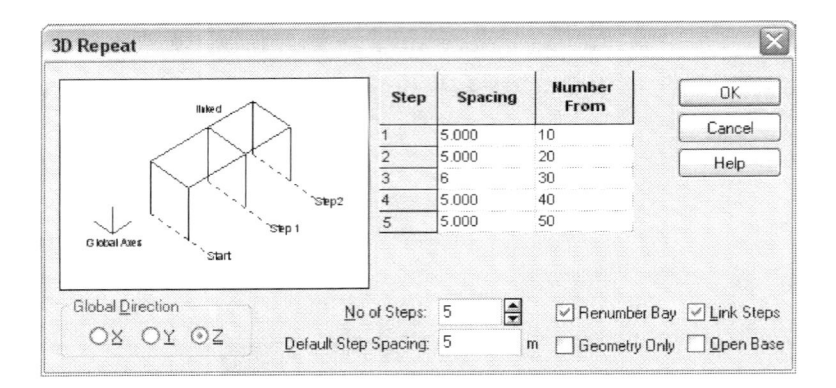

- Specify the **Global Direction**; you have three choices X, Y, or Z.
- Specify the **No. of Steps** excluding the geometry you draw.
- Specify the **Default Step Spacing** in the default length unit.
  - The Step Spacing may be positive or negative value (that is if the duplication process to take place in the negative side of X, Y, or Z).
  - You can change the individual step spacing from the table in the dialog box.
- Specify if you want to **Renumber Bay**s, a new column will be added, so user can specify the starting number of Beam numbers STAAD.*Pro* will start with, for each new frame will be added.
- Specify if you want to **Link Steps** or not. Linking Steps is to link the duplicate frames generated by Beams parallel to the direction of copying. Accordingly specify if you want to make the **Base** (the nodes at the bottom) to be linked or unlinked (**Open**).

Using Translational Repeat

Exercise 11

1. Start STAAD.*Pro*, and create a new frame 4 meter in X-axis, and 3 meter in Y-axis.

2. Save this file under the name **Common.std** (this file will be used for the other exercises of this module and the other modules).

3. Select all Beams.

4. Start the **Translation Repeat**, and specify the input data in a way to produce the structure as shown below (leave the **Spacing** to be 5m):

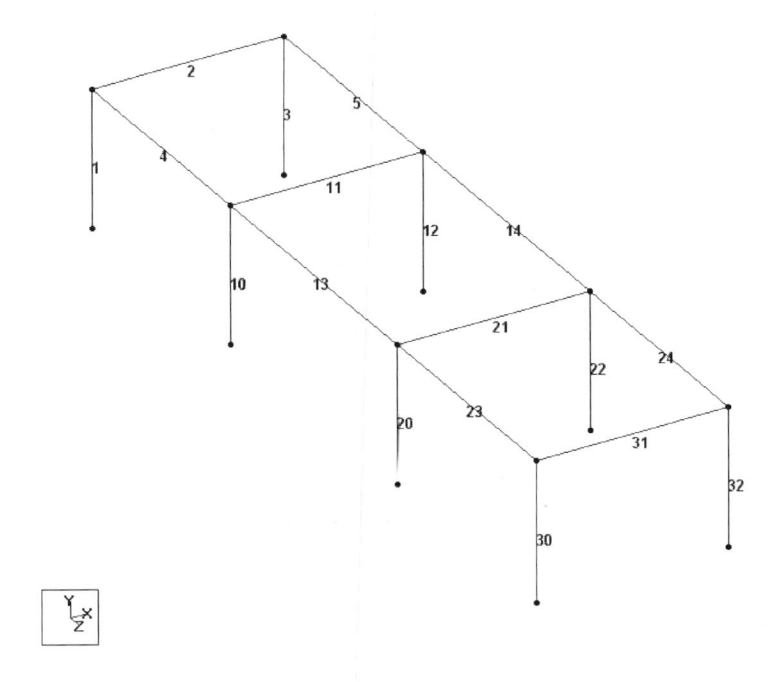

5. Don't save and close.

Circular Repeat

- With this function, we can duplicate Nodes, Beams, and Plates in a semi-circular, or circular fashion around one of the major axes.
- Select the desired Nodes, Beams, or Plates to be duplicated. From **Generate** toolbar, select **Circular Repeat**, or from menus select **Geometry/Circular Repeat**, the following dialog box will appear:

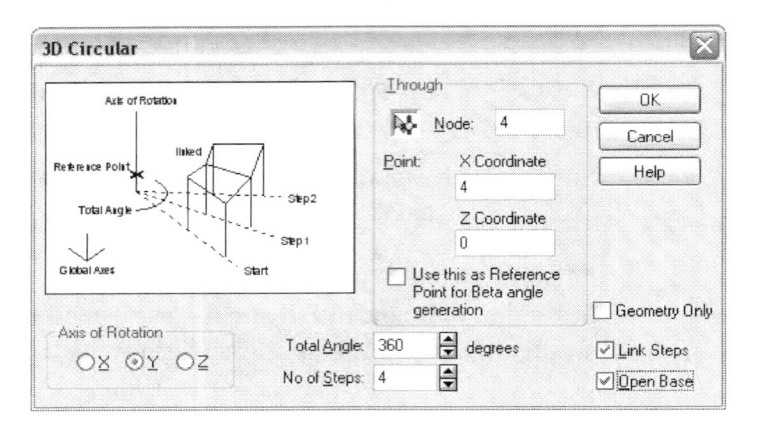

- Specify the **Axis of Rotation**, which will be one of the three major axes X, Y, or Z.
- Specify the **Total Angle** (+ve=CCW) to be covered by the duplicate frames. Then specify the **No. of Steps** excluding the geometry you draw.
- To specify the point that the Axis of Rotation will go **Through**, you have three ways:
  - Click on the icon, and specify it on the screen.
  - You remember the Node Number, type it in.
  - You don't know the Node Number but you know its coordinate, type it in.
- Specify to **Link Steps**, or not, and to **Open Base**, or not.

Using Circular Repeat

Exercise 12

1. Open Common.std.
2. Select all Beams.
3. Start the **Circular Repeat**, and specify the input data in a way to produce the structure below (this is a top view):

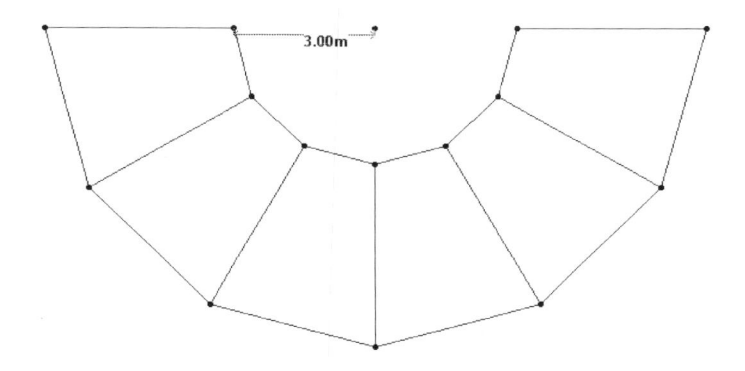

4. Don't save and close.

Mirror

- With Mirror function, we can create a mirror image of the selected Nodes, Beams, and Plates around any of the three planes.
- Select the desired Nodes, Beams, or Plates to be duplicated. From **Generate** toolbar, select **Generate-Mirror**, or from menus select **Geometry/Mirror**, the following dialog box will appear:

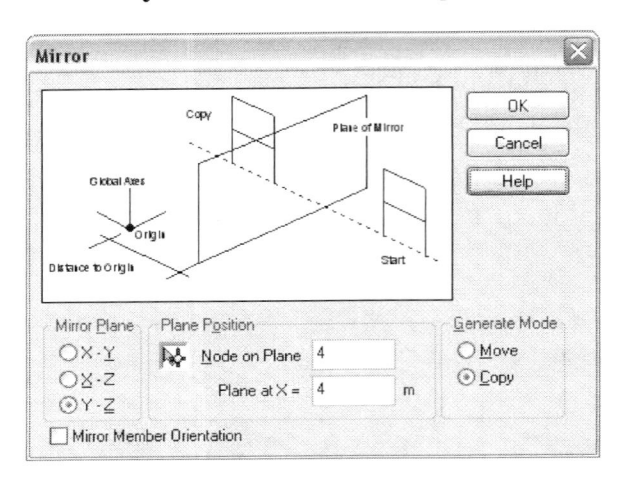

- Specify the **Mirror Plane** one of the following X-Y, X-Z, or Y-Z.

- To specify the **Plane Position**, you have three ways:
  - Click on the icon, and specify it on the screen.
  - If you remember the Node Number, type it in.
  - If you don't know the Node Number but you know it's X coordinate, type it in.

- Specify **Generate Mode** whether:
  - **Copy** mode will generate the mirror image and keep the original geometry.
  - **Move** mode will generate the mirror image and erase the original geometry.

Using Mirror

Exercise 13

1. Open Common.std.
2. Select all Beams.
3. Start the **Generate-Mirror**, and specify the input data in a way to produce the structure below.

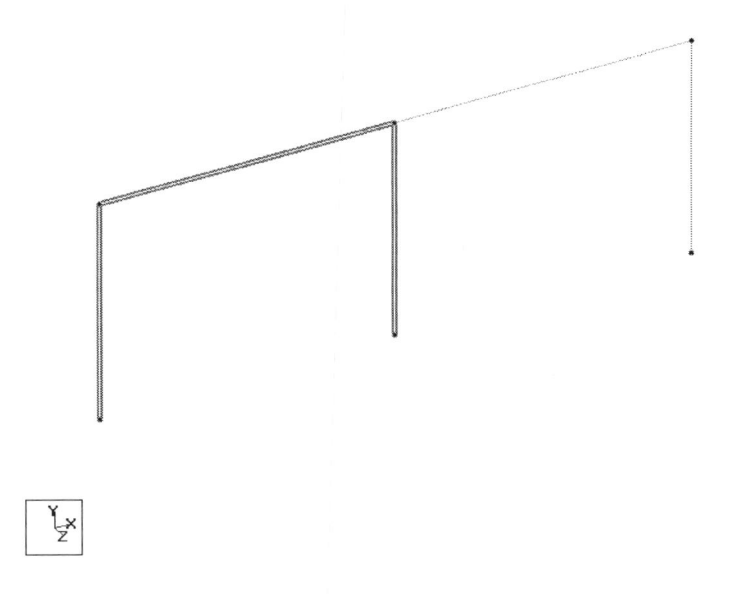

4. Don't save and close.

Rotate

- With Rotate function, we will be able to rotate Nodes, Beams, and Plates around any axis we specify.

- Select the desired Nodes, Beams, or Plates to be rotated. From **Generate** toolbar, select **Generate-Rotate**, or from menus select **Geometry/Rotate**, the following dialog box will appear:

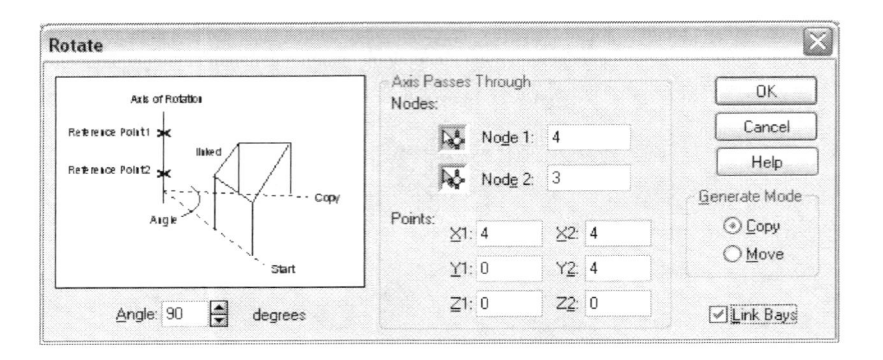

- Specify the rotation **Angle** in degrees.

- To specify the **Axis Passes Through** point you have three ways to do that:
  - Click on the icon, and specify Node 1, and Node 2 on the screen.
  - If you remember the Node Number, type it in.
  - If you don't know the two Nodes Number but you know their coordinates, type it in.

- Specify **Generate Mode** whether:
  - **Copy** mode will generate the rotated geometry and keep the original geometry. Specify to **Link Bays** or not.
  - **Move** mode will generate the rotated geometry and erase the original geometry.

Note
- Positive rotation angle will rotate Counter Clock Wise.

Using Rotate

Exercise 14

1. Open Common.std.
2. Select all Beams.
3. Start the **Generate-Rotate**, and specify the input data in a way to produce the structure below. (Hint: Use Rotate with angle =+40).

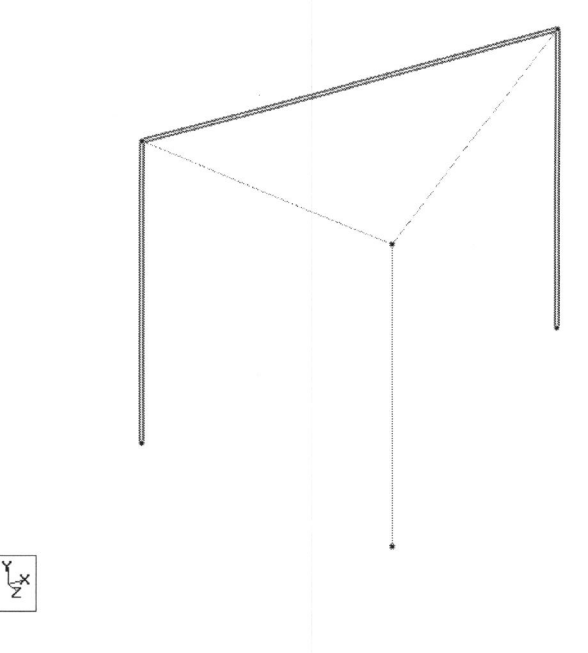

4. Don't save and close.

Move

- In Module 2, we found that STAAD.*Pro* geometry is the *Nodes*. Beams, and Plates are defined based on the Nodes at their ends or corners, hence when you move Nodes, as if you are moving Beams, or Plates, or stretching them (stretch here means the two movement; elongating or shortning)
- Select the desired Nodes (you can select Beams, or Plates, but move will move their Nodes).
- You have three ways to access the function:
  - Press F2.
  - Right-click and select Move.
  - From the menus select **Geometry/Move**, then one of the options.
- The following dialog box will appear:

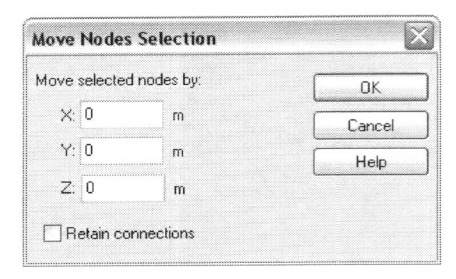

- Input the movement disctance and which direction.
- Depends on your selection the output will be either moving or stretching.

Insert Node

- Node in STAAD.*Pro* is stiffed joint, hence there will be 6 reactions on it. That means, if user inserted a Node, at the middle of a Beam, the stability of the structure will not be affecetd.
- Select one Beam.
- You have three ways to access the function:

  - Start the **Insert Node** function by clicking on the function from the **Geometry** toolbar
  - Select **Geometry/Insert Node**, or **Geometry/Split Beam**!
  - Right-click and select **Insert Node**.

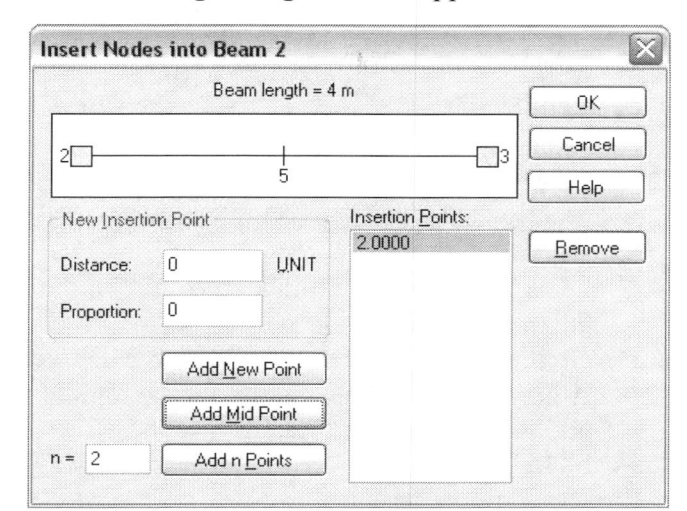

- The following dialog box will appear:

Add Mid Point
- If you click this button, a new Node at the middle of the Beam will be added.

Add n Points
- In the field of **n** specify the number of Nodes to be added, then click **Add n Points**.

Add New Point
- In the field of **Distance** specify the location from the start of the Beam, and click **Add New Point**. STAAD.*Pro* will produce in the field of **Proportion**; the percentage of the diatnace to the whole length of the Beam.
- Alternatively, you can input the **Proportion**, and the **Distance** will be measured.

Note
- If you select more than one Beam and initite the function, the following dialog box will appear:

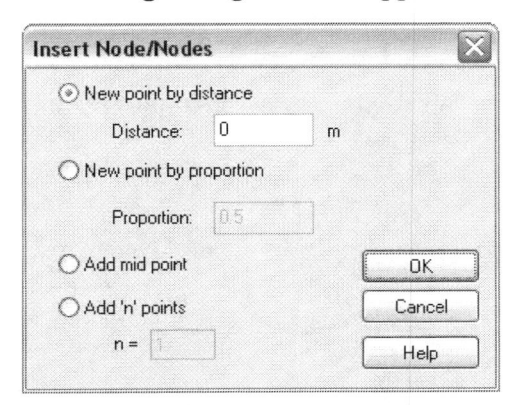

- This dialog box will give you the ability to perform this function on multi-beams in one single command.

Add Beam between Mid-Points

- To add Beams from the middle of one Beam to the middle of another Beam.
- From the Geometry toolbar, click the **Add Beam between Mid-Points**.
- The mouse shape will change to this shape.
- Click the first Beam (click anywhere in the Beam), a new Node at the middle will be added.
- Go the second Beam, and click anywhere on the Beam.
- A new Beam will be added fom the middle of the first Beam to the middle of the seocnd Beam.

Add Beam by Perpendicular Intersection

- The same as the previous function except this function will link an existing Node in a perpendicular fashion to an existing Beam.
- From the Geometry toolbar, click the **Add Beam by Perpendicular Intersection**.
- Click on an existing Node.
- Click on any Beam.
- A new Beam will be added perpendicular on the selected Beam.

Module 3: Useful Functions to Complete the Geometry

Using Tools to add Nodes and Beams

Exercise 15

1. Open Common.std.
2. Select the 4 m Beam.
3. Using Insert Node, and Move try to do the geometry below:

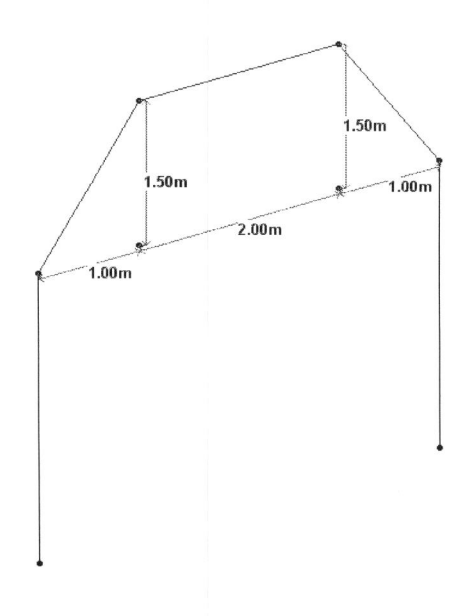

4. Do the following steps to produce the below geometry from the geometry you just created:

   a. Add Beams

   b. Add Beam between Mid-Points

   c. Add Beam by Perpendicular Intersection

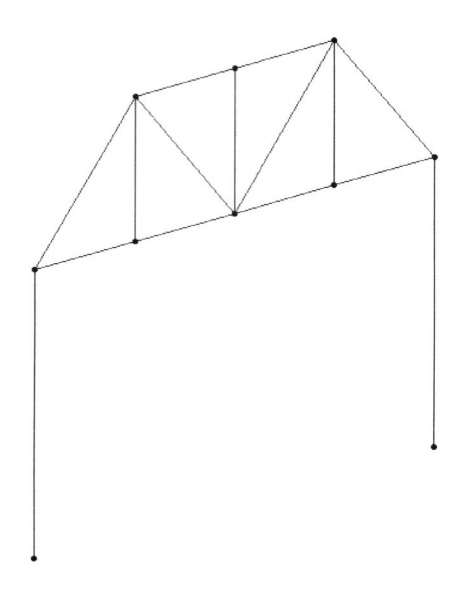

Connect Beams along an Axis

- You can connect Nodes along any of the major axes with Beams.
- Select the desired Nodes to be connected along X, Y, or Z.
- From the menus select **Geometry/Connect Beams along**, then choose one of the three axes. The Nodes will be connected with Beams.

Intersect Selected Members

- Sometimes, it happens that two Beams intersect each other without creating a Node at the interesction point. This may happen specially in DXF files importing. This will lead STAAD.*Pro* to interpret that there will be no transmitting of the forces between these two Beams, therefore we need to make a check for such a case, and correct it.
- From the menus select Geometry/Intersect Selected Members, then you can select Intersect, or Highlight.
  - Highlight will only highlight any occurance of such a problem.
  - Intersect will solve the problem.
- A dialog box will appear asking you to specify the tolerance. Specify the tolerance value, click OK.

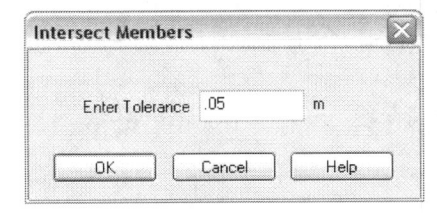

- If such a problem is not present in your geometry, STAAD.*Pro* will produce the following message:

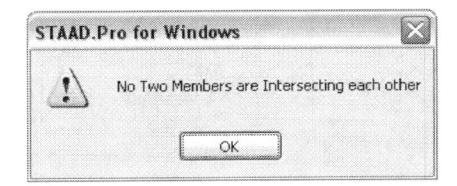

- If such a problem is present in your geometry, STAAD.*Pro* will create Nodes at intersection as needed, and produce the following message:

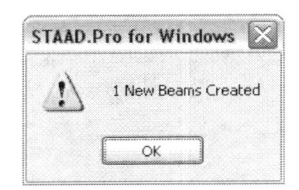

Note
- If you don't want to use **Highlight** function, select the beams you suspect have this problem, then issue the command **Intersect**.

Connecting Nodes, and Creating Intersections

Exercise 16

1. Open Common.std.
2. Select all Beams, and start Translational Repeat. The Global Direction is Z, Default Step Spacing is 3m, and No. of Steps is 3, **(Don't Link Steps)**, Click **OK**.
3. The shape should look like the following:

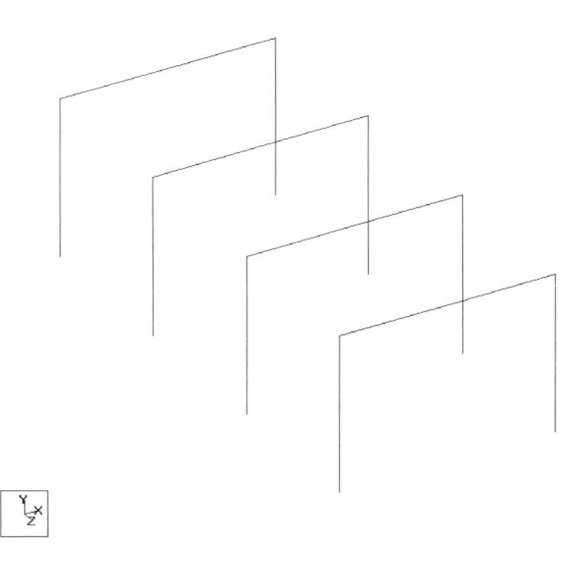

3-18

4. Change to the Node cursor, and select the Nodes at the upper-right, like the following:

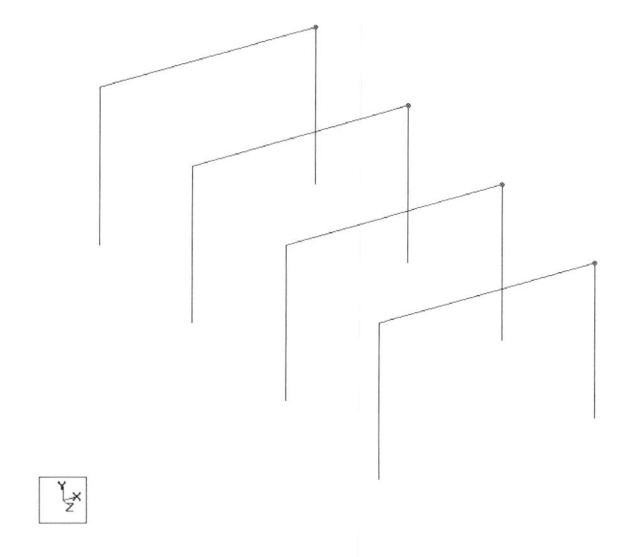

5. From Menus select **Geometry/Connect Beams Along/Z Axis**. The shape should be like the following:

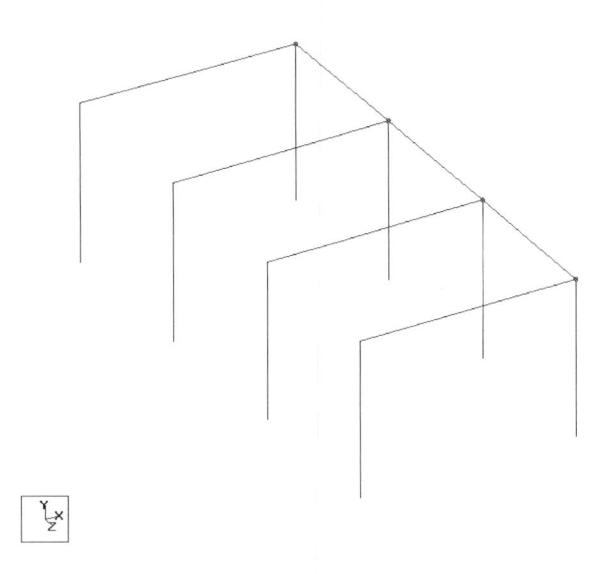

6. Select the first Node and the last Node, on the upper-left side:

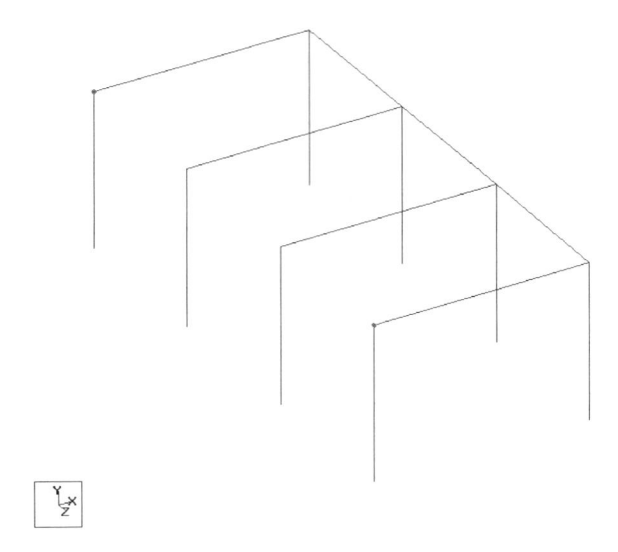

7. From Menus select **Geometry/Connect Beams Along/Z Axis**. A single Beam will be added connecting the first Node and the last Node.

8. Change to the Beam cursor.

9. Select the whole geometry (Ctrl+A).

10. From menus select **Geometry/Intersect Selected Members /Intersect**, and accept the default tolerance value.

11. A dialog box will tell you that **"2 new Beams Created"**.

12. Don't save and close.

Cut Section

- Dealing with 3D View of a geometry with all of the Beams and Plates shown may lead to confusion, and accordingly will slow the production of geomerty or other functions like inputting cross-sections. Cut Section is the suitable function to create a slice of the geometry so user can focus on the job.
- From **Strcuture** toolbar, select **Cut Section**, or from menus select **Tools/Cut Section**.

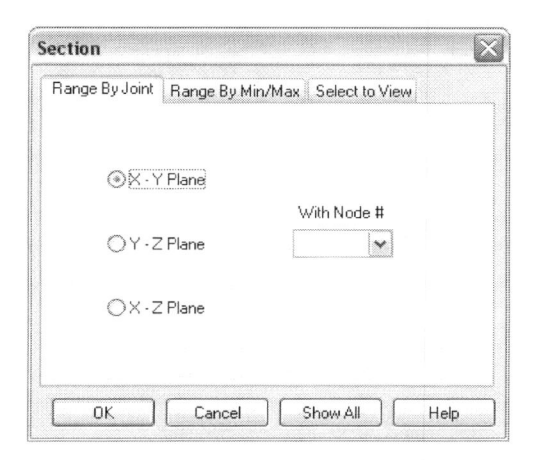

- You have three ways to create a slice of your geometry:
  - Range By Joint
  - Range By Min/Max
  - Select to View

Range By Joints
- Specify the Plane you want to slice parallel to it. You have three choices, X-Y, Y-Z, and X-Z.
- Specify a Node number, so the slicing will take place at it.

Range By Min/Max
- Specify the Plane you want to slice parallel to it. You have three choices, X-Y, Y-Z, and X-Z.
- Specify the **Minimum** distance, and the **Maximum** distance, any geometry parallel to the plane selected and with the range of the distance will be shown.

[Section dialog: Range By Min/Max tab with X-Y Plane, Y-Z Plane (selected), X-Z Plane options and Minimum/Maximum fields]

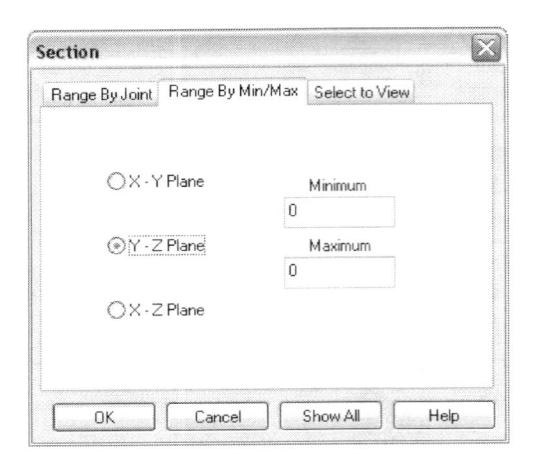

Select to View ■ You have three choices in this slicing method:

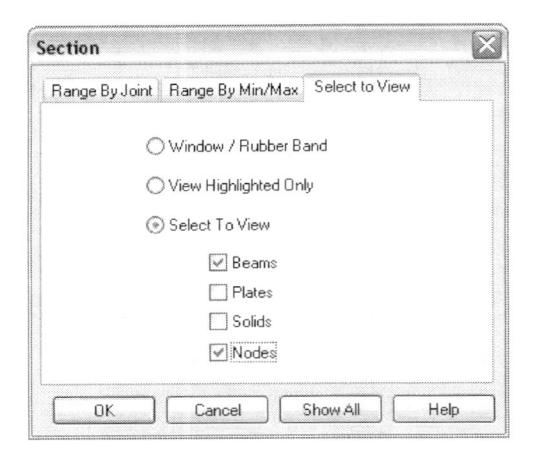

[Section dialog: Select to View tab with Window/Rubber Band, View Highlighted Only, Select To View (selected) with Beams, Plates, Solids, Nodes checkboxes; Beams and Nodes checked]

- **Window/Rubber Band**; whereis the user to make a window (click left button and hold to make a rectangle) whatever in it will be shown. (Note that the *Middle Point* of the member is the important part to be included inside the Window, and any Node within the Window).

- Select Nodes, Beams, or Plates, then choose **View Highlight**. If you selected Nodes, Nodes only will be shown, etc.

- Click on **Select To View**, and choose the part of geometry you want to see, you can select more than one choice.

■ According to your choice, part of the structure will be shown.

Show All ■ Whenever you are done working with your desired slice, simply select the function again, and click on **Show All** button, the whole structure will appear again.

Renumber

- Numbering Nodes, Beams, and Plates is the mission of STAAD.*Pro*, but we can interfere in it and renumber whatever we want.
- Select Nodes, Beams, or Plates.
- From menus, select **Geometry/Renumber**, choose suitable option.
- Warning message will appear:

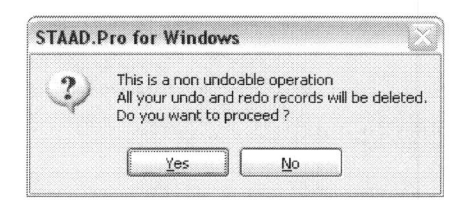

- Read this message carefully, as it will remove any undoing from this file, hence it may be something you don't want to do.

- If you click on **No**, nothing will happen. If you click **Yes**, a dialog box will appear:

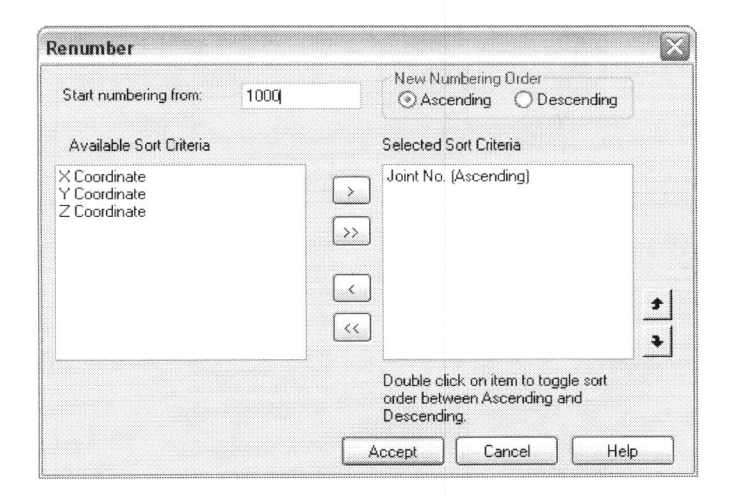

- Specify **Start numbering from**, and whether it will in **Ascending** order, or **Descending** order, then specify the **Sort Criteria**.

Cutting Section & Renumbering

Exercise 17

1. Open Common.std.

2. Select all Beams, and start Translational Repeat. The Global Direction is Z, Default Step Spacing is 3m, And No. of Steps is 3, Link Steps, and Open Base.

3. Right-click and select **Labels**, and then click **Node Numbers** on. Identify one of the Nodes at the top (let it be Node # 2).

4. Start **Cut Section**, at the **Range By Joint**, select **X-Z**, **With Node #** 2. The shape of the Geometry should look something like this:

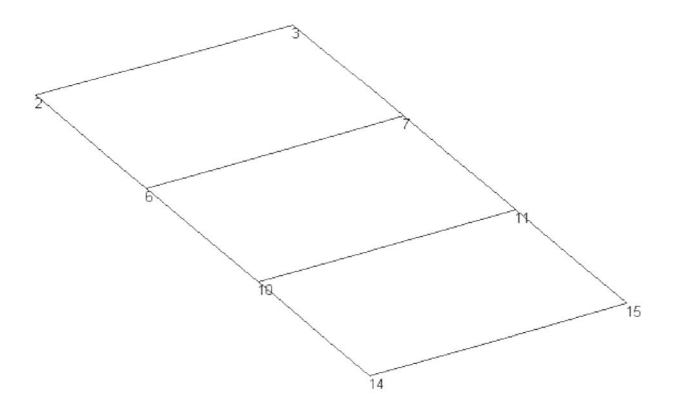

3-24

Module 3: Useful Functions to Complete the Geometry

5. Change the cursor to **Node** cursor, and select all Nodes.
6. Select **Geometry/Renumber/Nodes**.
7. Select to start numbering from 100, and in an **Ascending** order. Specify the Sort Criteria to be **Joint No.**
8. The geometry should look like the following:

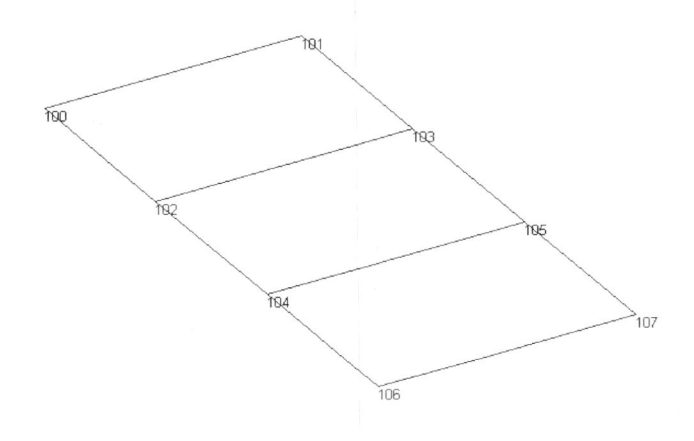

9. Select **Cut Section** again, and click **Show All**.
10. Don't save and close.

Delete

- You can delete Nodes, Members, and Plates.

Deleting Nodes
- If you want to delete Nodes, any thing attached to them (namely; Beams ot Plates) will be deleted:
  - Select Nodes.
  - Press **Del** Key at the keyboard.
  - A warning message will appear:

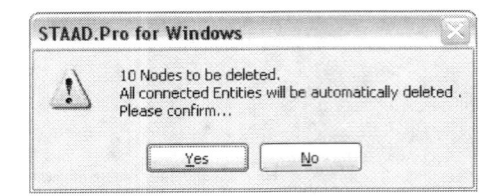

  > STAAD.Pro for Windows
  > 10 Nodes to be deleted.
  > All connected Entities will be automatically deleted.
  > Please confirm...
  > [Yes] [No]

  - If you click **Yes**, the operation will be done and the Nodes will be deleted. **No**, will stop the operation.

Deleting Beams and Plates
- Select the desired Beams, and Plates:
  - If you select one Beam and press **Delete**, the Nodes at its end will remian intact (except if your geomerty is one Beam).
  - If you select more than one Beam, STAAD.*Pro* will give you a warning message:

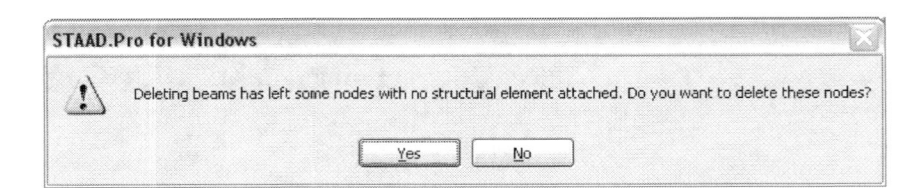

  > STAAD.Pro for Windows
  > Deleting beams has left some nodes with no structural element attached. Do you want to delete these nodes?
  > [Yes] [No]

  - Based on your answer, you can keep the Nodes, or remove them as well.
- The same thing applies for Plates, except deleting a single Plate will lead to the warning message.

Module 3: Useful Functions to Complete the Geometry

Undo / Redo

- You have unlimited number of Undos, and Redos to perform in one session.
- After you perform several functions, you want to undo one or two of them, from the File toolbar, click **Undo**, or select **Edit/Undo**, or you can press **Ctrl+Z**.
- Also, you can undo group of actions, if you clicked on the pop-up list, you can click on a certain action, accordingly STAAD.*Pro* will highlight anything between what we select and the point you are standing on it right now, then double-click or press Enter.

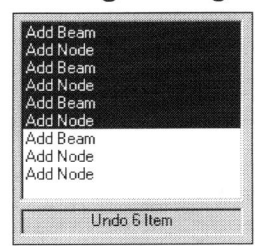

- Redo is to undo the undone
- From **File** toolbar, click **Redo**, or from menus select **Edit/Redo**, or press **Ctrl+Y**.
- All Undo specifications apply to Redo.

Zooming and Panning

- You can zoom and Pan using different methods in STAAD.*Pro*. You will find all zoom commands and Pan command in **View** toolbar, and from menus in the **View/Zoom** or **View/Pan**.
- **Zoom Window**, is to draw a window (rectangular shape by specifying two opposite corners) around the desired portion of the geometry to be magnified, by clicking the left button of mouse and holding until you specify the other corner of the window. Whatever inside this window will be magnified.
- **Zoom Factor**, is to specify a magnifying factor (greater than 1) or shrinking factor (less than 1), the following dialog box will appear:

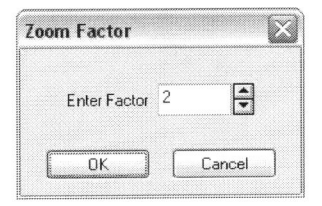

3-27

- **Zoom In**, to get you closer to the geometry, step-by-step.
- **Zoom Out**, to get you farther from the geometry, step-by-step.
- **Zoom Extents** (the menu option is All), after several zooms in and out, this options allows you to see the whole geometry filling the screen.
- **Zoom Dynamic**, just like **Zoom Window**, except Zoom Dynamic will create a window conatining the new view, and you can scroll up and down, left and right, to view the other parts of the structure.
- **Zoom Previous**, to get you back to last view.
- **Pan**, a hand will appear, to allow the user to pan the geometry in all of the sides, click the left button and drag the geometry in any direction (it is more practical to use Pan with other zooms), to disabe it press **Esc**, or click the icon again.
- **Display Whole Structure**, just like **Zoom Extents**. It works with Cut Section (discussed earlier) as **Show All** option.
- **Magnifying Glass**, a glass will appear, click and hold, you will see a bigger picture of the part, once you start moving you will magnify other parts of your geometry.

Using IntelliMouse
- If you have IntelliMouse (the mouse with a wheel), you can use it to Zoom In, and Zoom Out.
  - If you move the wheel forward you are zooming in.
  - If you move the wheel backward you are zooming out.
- **Previous Selection**, STAAD.*Pro* remembers the last selection you made. To re-use the last selection, simply click this button, and STAAD.*Pro* will select it for you.

Dimensioning

- You can dimenstion your geometry to make it more meaningful for anybody will view your model.
- There are two ways to put dimension in STAAD.*Pro*:
  - Using **Dimension** function
  - Using **Node to Node Distance** function

Dimension function

- Dimesion function will put dimension over Beams only, stating the length of the Beam on the middle of it.
- From Structure toolbar, click **Dimension**, or from menus select **Tools/Dimension Beams**. The following dialog box will appear:

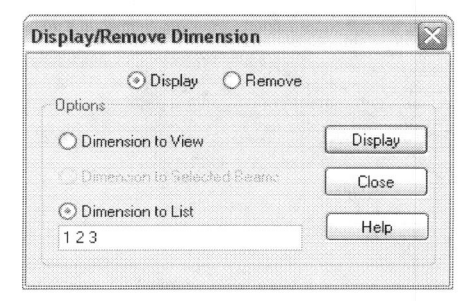

- While Display mode is on, select either to:
  - **Dimension to View**, to dimension all Beams at the current view.
  - **Dimension to Selected Beams**, the selected Beams prior to the initiation of this function will be dimensioned.
  - **Dimension to List**, type in the numbers of Beams that you desire to dimension (leave spaces between Beam numbers).
- To remove the dimension from the geometry, select the **Remove** mode, and click **Remove**.

Node to Node Distance function

- This function is manual, you will click on two Nodes and STAAD.*Pro* will put a dimension between these two Nodes. There is no need for a Beam to be between the two Nodes.
- From the Structure toolbar, click the Node to Node Distance, or from menus select Tools/Display Node to Node Dimension.
- In both ways, the mouse shape will change.
- Click the first Node, then click the second Node, dimension will appear on the distance you clicked. Keep doing this, until you are done.

- To remove this type of dimension, from **Structure** toolbar, click **Remove Node to Node Distance**, or from menus select **Tools/Remove Node Dimension**. The dimension will disappear.

Pointing to Nodes, Beams, and Plates

- Without issuing any command, if you point to Node, Beam, or Plate STAAD.*Pro* will give you the number of the Node, Beam, or Plate.

Pointing to Node
- Using the Node cursor, point to a Node, the cusor will provide the Node Number:

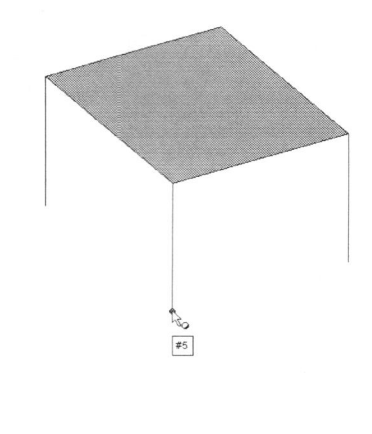

Pointing to Beam
- Using the Beam cursor, point to a Beam, the cursor will provide the Beam Number, and the Start and End Colors:

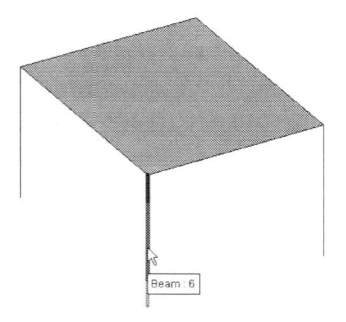

Module 3: Useful Functions to Complete the Geometry

- As we mentioned before, each Beam is defined by the two Nodes at its ends. Since STAAD.*Pro* will number every thing in the geometry, we will never know how STAAD.*Pro* wrote the definition of Beam in the input file except using the following:
  - Opening the input file, and verify.
  - Go the the Beam table in the **Geometry** Page Control.
  - Using the color code.
- By default the green will be the Start, and blue is the End.
- This particularly important when you want to use Loading system.

- To change the colors of the Start and End:
  - Right-click, and select **Labels**, from **Beams**, there will be two colors labeled **Start Color**, and **End Color**. Click the desired color icon, and change the colors.

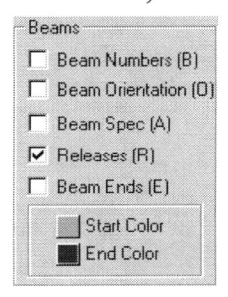

- To permanently set the color code of the Beams, click **Beam Ends** on in the same dialog box.

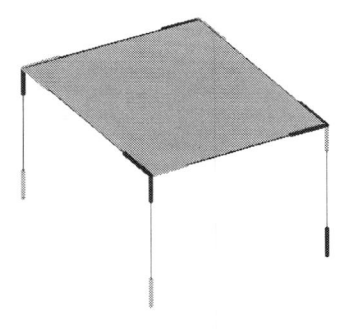

Pointing to Plate ■ Using the Plate cursor, point to a Plate, the cursor will provide the Plate Number.

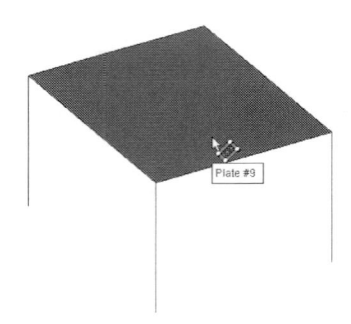

Global and Local Coordinate Systems

- There is a single **Global Coordinate System** in STAAD.*Pro*, which we defined the Node coordinates using it.
- The **Global Coordinate System**, appear at the lower left corner of the main window.
- We use GX, GY, and GZ, to differntiate them from the Local Coordinate System X, Y, and Z.
- For each Beam, there is a Local Coordinate System, as follows:
    - X, is always from Start to End along the member.
    - Y, in the direction of Minor principle axis.
    - Z, in the direction of Major principle axis.

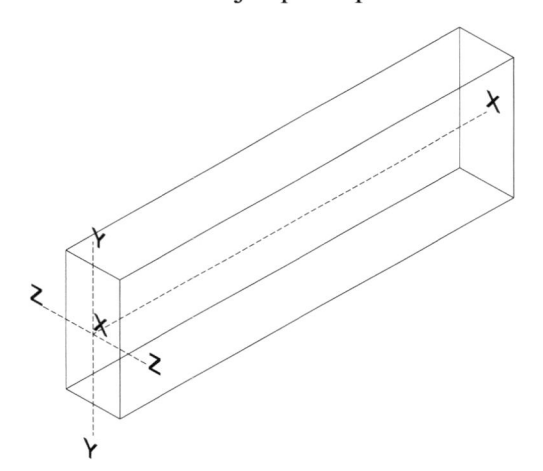

- Beam results always produced using the Local Coordinate System, such as Fx (Axial Load), Fy (Shear), and Mz (Bending moment).

- To view the Local Coordinate System of the Beam
    - Right-click, and select **Labels**.
    - From **Beams**, click **Beam Orientation** ON

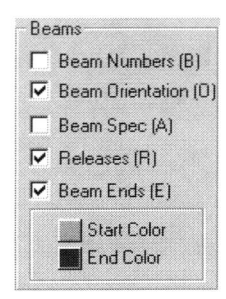

- You will see the X, and Y, and you can determine Z.

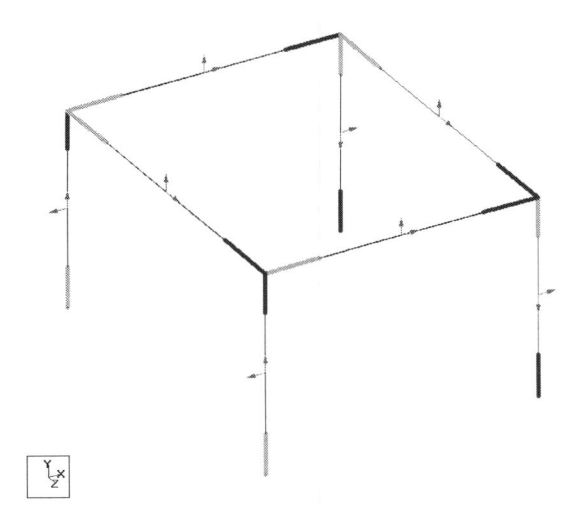

- For each Plate there is Local Coordinate System, which will be as follows:
    - X, is from first Node to the second Node.
    - Y, lies in the Plane defined by the third point orthogrphical to X.
    - Z, is derived from the Right-Hand-Rule (which defines the relationship between the three axes).
- To see the orientation of Plate Local Coordinates:
    - Right-click, and select **Labels**.
    - From **Plates**, click **Plate Orientation** ON. 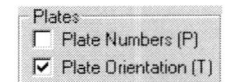
- You will see the X, Y, and Z of the Plate:

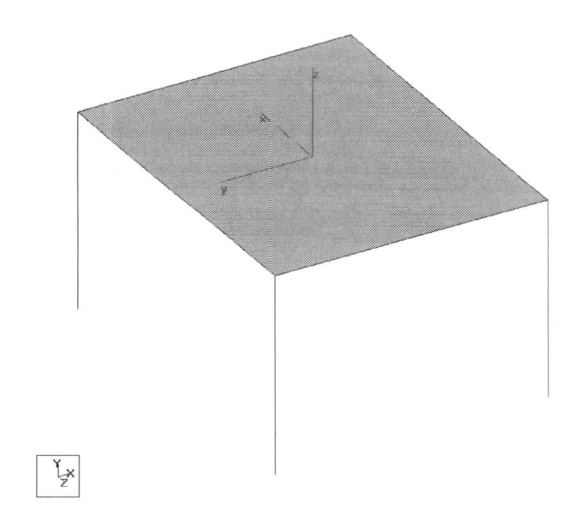

Module Review

1. In the Mirror function which of the following statements is true:
    a. I can define a User-Defined Plane to Mirror around it.
    b. I can only Mirror around X-Z Plane.
    c. I can Mirror around any of the three main Planes.
    d. There is only Copy mode in Mirror.

2. In the Translational Repeat:
    a. The distances between steps generated should be the same.
    b. I can repeat parallel to any of the three main axes.
    c. I can't link steps.
    d. I can't renumber step Beams.

3. Use _____ command to cut a slice of your geometry.

4. If you delete Nodes, the Beams, and Plates attached to them will be intact:
    a. True
    b. False

5. There is one method to place dimension on your geometry:
    a. True
    b. False

6. Some of the Zooms options are _____,_____,_____,_____

Module Review Answers

1. c
2. b
3. Cut Section
4. b
5. b
6. Window, In, Out, Dynamic, Previous, etc.

Module 4:

Properties

This module contains:

- Prismatic Property Type
- Built-In Steel Table Property Type
- Thickness Property Type
- Miscellaneous Functions

Introduction

- After the creation of the geometry, next step will be to Assign Properties (Cross-Section) to each Beam, and Plate.
- In our discussion we will select Beams and Plates prior to initiating the commands. This method is easier and handier especially if you want to assign to big number of, Beams, and Plates.
- All of work in this module will be in the **General** Page Control.

Property Types

- As we mentioned in Module 1, STAAD.*Pro* needs the cross-section to calculate I (moment of inertia), hence calculate K (stiffness factor), and without K STAAD.*Pro* cannot complete the system of matrices mentioned before and accordingly cannot produce the needed results. This will show the importance of this part to the whole process.
- There are two main methods to assign a property (cross-section) to Beams, and one method to assign a property (thickness) to Plates.
- For Beams, the two main methods are:
  - Prismatic, for Concrete sections
  - Built-In Steel Section Table and Built-In Aluminum Section Table.
- For Plates, the only available method is Thickness.

STAAD.*Pro* 2005 Tutorial

Type 1: Prismatic

- Go to **General** Page Control.
- Make sure you are in the **Property** sub-page.
- Select the desired Beams (Beams or Columns). 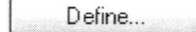
- From Data Area at the right of the screen, click **Define**.
- There will be four different cross-sections to pick from:

Circle
- Specify the Diameter of the cross-section:

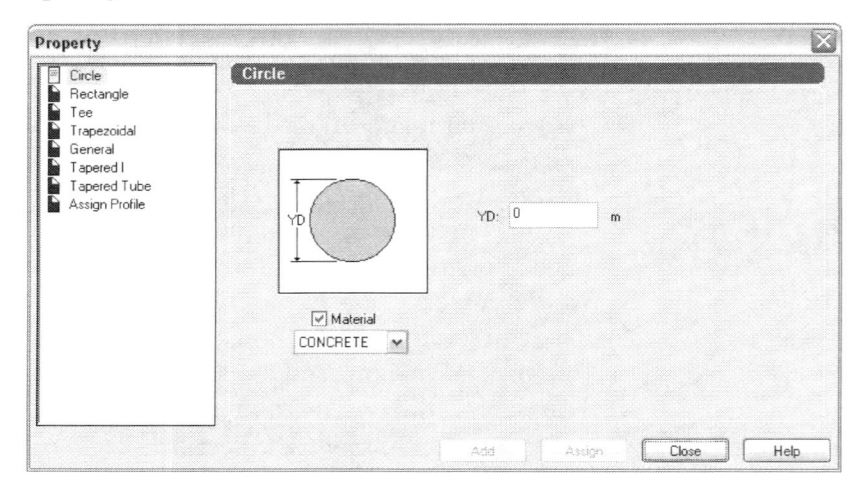

Rectangle
- Specify the Depth (YD-means Depth parallel to local Y) and Width (ZD-means Depth parallel to local Z) of the cross-section.

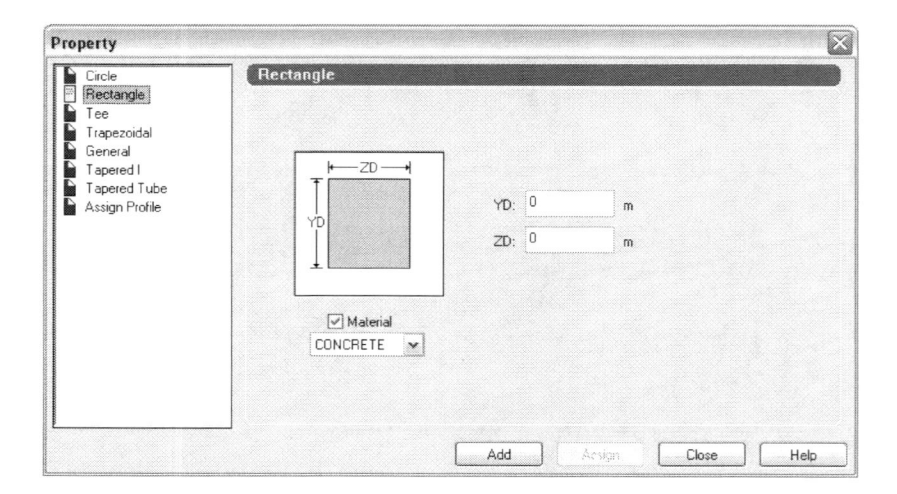

4-4

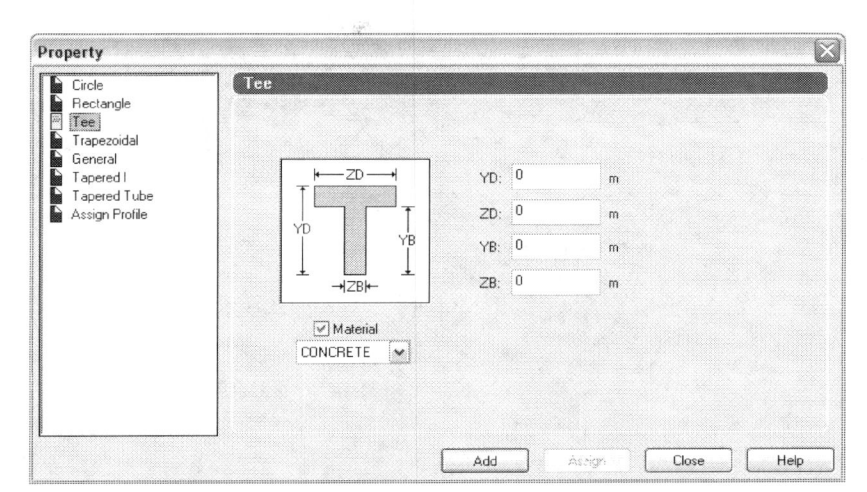

Tee ■ Specify the four types of data required.

Trapezoidal ■ Specify the three types of data required.

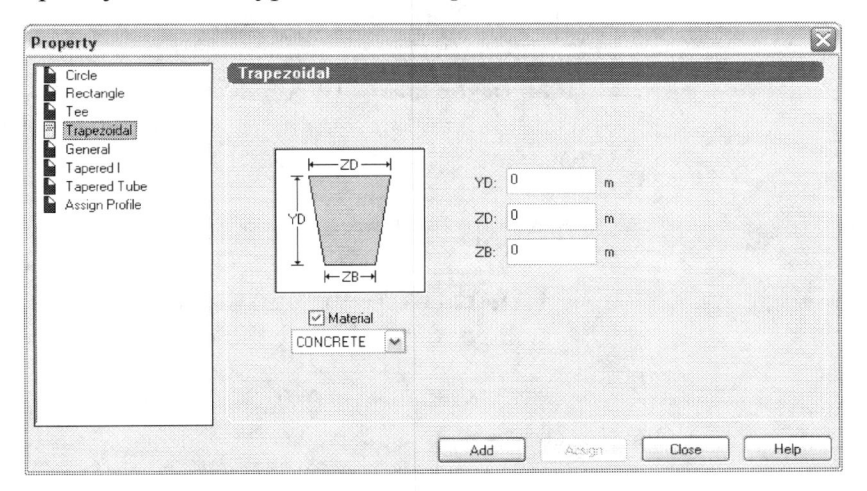

- For the above four cross-sections, STAAD.*Pro* will take the values supplied by the user and calculate all the needed information, which they are:
  - Ax, Cross –Sectional Area
  - Ay, Effective Shear Area for shear forces parallel to local Y-axis
  - Az, Effective Shear Area for shear forces parallel to local Z-axis
  - Ix, Moment of Inertia about X-axis (Torsional Constant)
  - Iy, Moment of Inertia about Y-axis
  - Iz, Moment of Inertia about Z-axis

General ■ As an alternate method, you can give STAAD.*Pro* all the data needed (Ax, Ix, Iy, and Iz) with or without the cross section of the Beam, by selecting **General**.

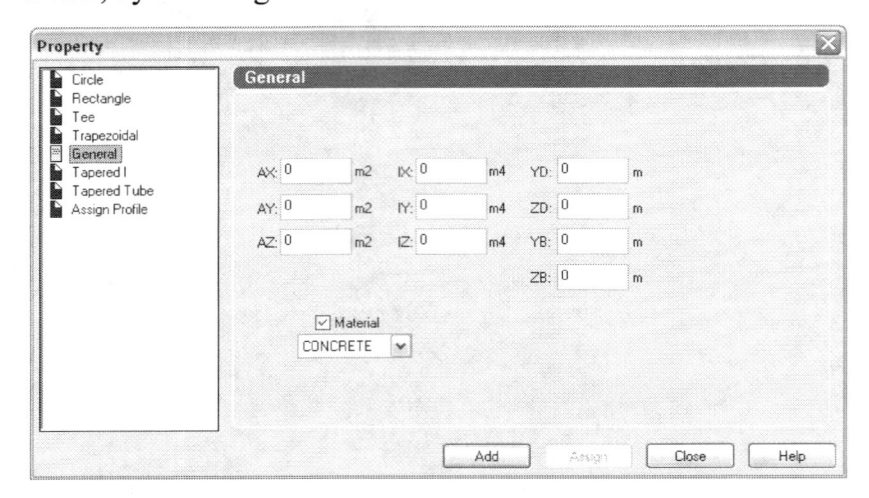

Note ■ Also, you can reach the same command from the menus by selecting **Commands/Member Property/Prismatic**.
■ In all of the above five dialog boxes you will find a check box called **Material**, and pop-up list with selection of **Concrete**. You have two ways to deal with this issue:

- Accept this check box, and hence you are accepting the data available at STAAD.*Pro* (to view the default material data, from **General** page, **Property** sub-page, click on **Material** button, you will get the following dialog box)

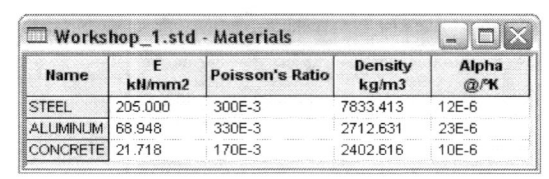

- Click this checkbox off, and hence you have to input the specific Material constants later on.

Module 4: Properties

Viewing Cross-Section

- To view cross section, right-click anywhere in STAAD window. From the menu select **Structure Diagrams**. From the dialog box, and under **3D Section**, select **Section Outline**.

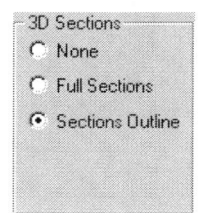

- The following will be produced:

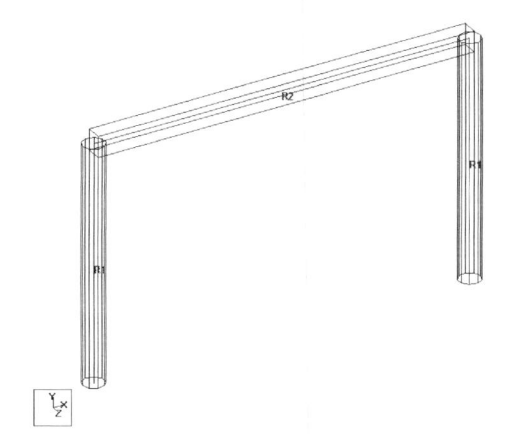

- From the same place, if you select **Full Section**, this is what you will get:

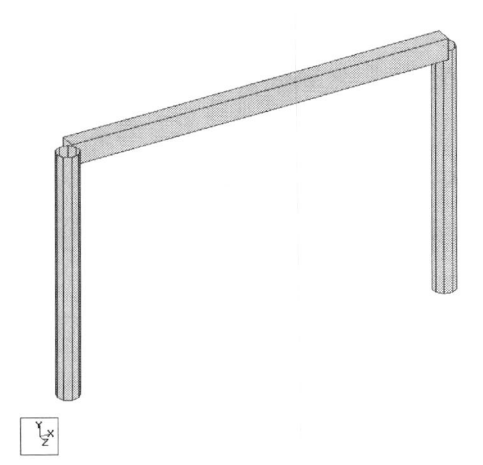

4-7

- From **View** toolbar, select **3D Rendered View**, or from menus select **View/3D Rendering**. Also, if you right-click at the STAAD.*Pro* window, you can select **3D Rendering**. In all cases this is what you will get:

Assign Prismatic Property to Beams

Exercise 18

1. Open Common.std
2. Assign for the two columns Circular shape YD=0.25 m
3. Assign for the beam Rectangular shape YD=0.35m, ZD=0.2m
4. View **Section Outline**, **Full Section**, and **3D Rendered View**.
5. Don't save and close.

Module 4: Properties

Type 2: Built-In Steel Table

- Go to **General** Page Control.
- Make sure you are in the **Property** sub-page.
- Select the desired Steel Beams.

Section Database

- From Data Area at the right of the screen, click **Section Database**.
- The following dialog box will appear:

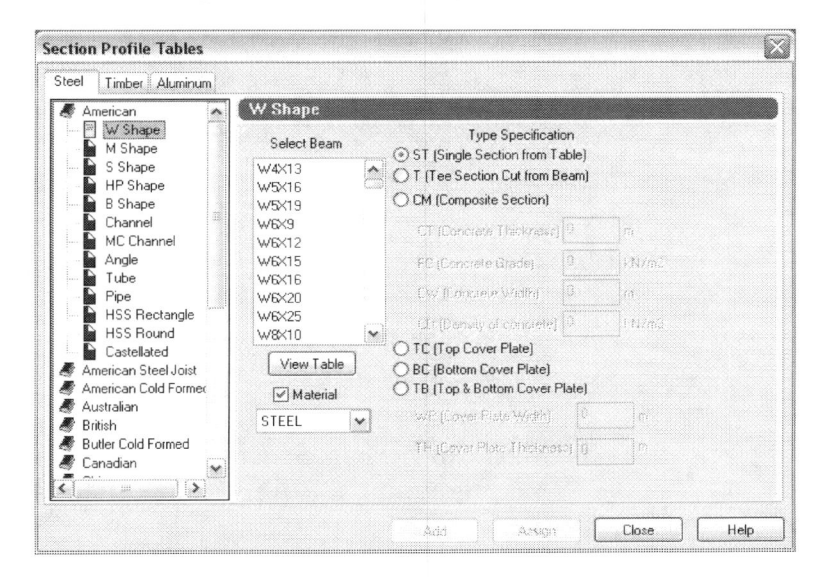

- By default The American table is displayed, you can choose other tables if you wish.
- Select the type: W Shape, M Shape, S Shape, etc.
- Select the size.
- Select the Specification.

Section Specification
- There are mainly five different section categories:
  - I-section (regardless of the naming convention of a specific table).
  - Angle.
  - Channel
  - Tube
  - Pipe
- For each category there will be specific specifications.

4-9

I-Section Specification

- **ST**, means standard section as mentioned in the table.
- **T**, means T-section cut from I-section.
- **CM**, means Composite Section—Steel and Concrete. Specify:
  - **CT**, Concrete Thickness
  - **CW**, Concrete Width
  - **FC**, Concrete Grade
  - **CD**, Density of Concrete

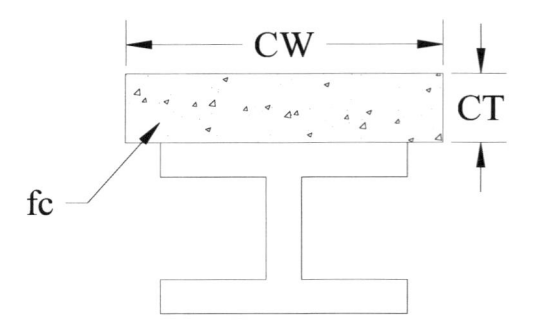

- **TC**, Top Cover Plate. Specify:
  - **WB**, the Width of Cover Plate
  - **TH**, the Thickness of Cover Plate

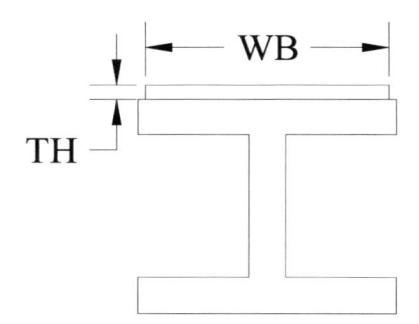

- **BC**, means Bottom Cover Plate. Specify the same data of TC.
- **TB**, means Top and Bottom Cover Plate. Specify the same data of TC.

Angle Specification
- **ST**, means standard section as mentioned in the table.
- **RA**, means an Angle with Reverse Y-Z:

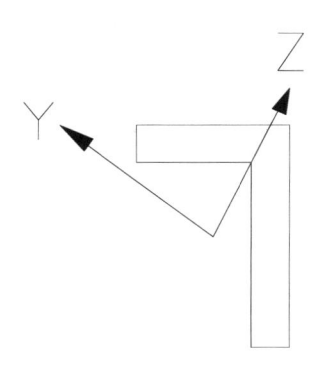

- **LD**, means Double Angle, Long Leg Back-to-Back. Specify:
  - **SP**, the space between the two Angles.

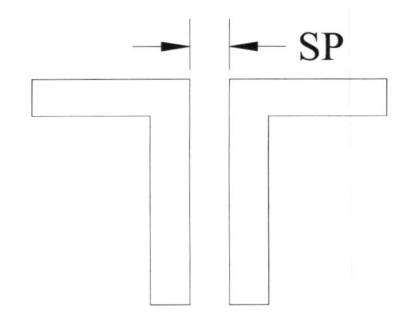

- **SD**, Double Angle, Short Leg Back-to-Back. Specify **SP** also.

Channel Specification
- **ST**, means standard section as mentioned in the table.
- **D**, means Double Channel Back-to-Back. Specify **SP**.

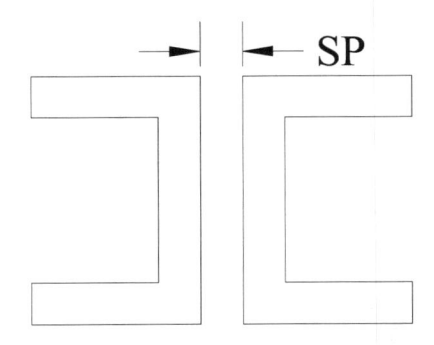

Tube Specification
- Either to select one of the pre-defined Tubes in the table.
- Or define your own. Specify:
  - **TH**, means the Thickness of the Tube
  - **WT**, means the Width of the Tube
  - **DT**, means the Depth of the Tube

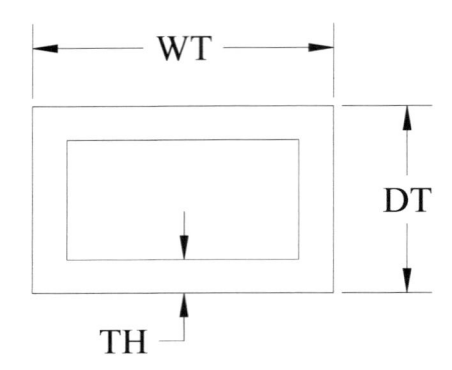

Pipe Specification
- Either to select one of the pre-defined Pipes in the table.
- Or define your own. Specify:
  - **OD**, means the Outside Diameter of the Pipe
  - **ID**, means the Inside Diameter of the Pipe

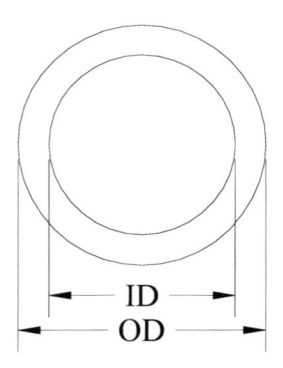

Note
- Also, you can reach the same command from the menus by selecting **Commands/Member Property/Steel Table** and then you can select the desired table.
- **Material** checkbox is the same as in Prismatic.

Module 4: Properties

Assign Built-In Steel Section to Beams

Exercise 19

1. Open Common.std.
2. Connect the diagonal Nodes by Beams to create bracing.
3. Select the two columns. From American Table, assign HP13X87
4. Select the beam. Assign W8X21
5. Select the Bracing members. Assign L40405 Double Angle Long Leg Back-To-Back with 0.05 m space between the two legs.
6. From **Strcuture Diagrams** set the **3D Section** to **Section Outline**. This is the end result:

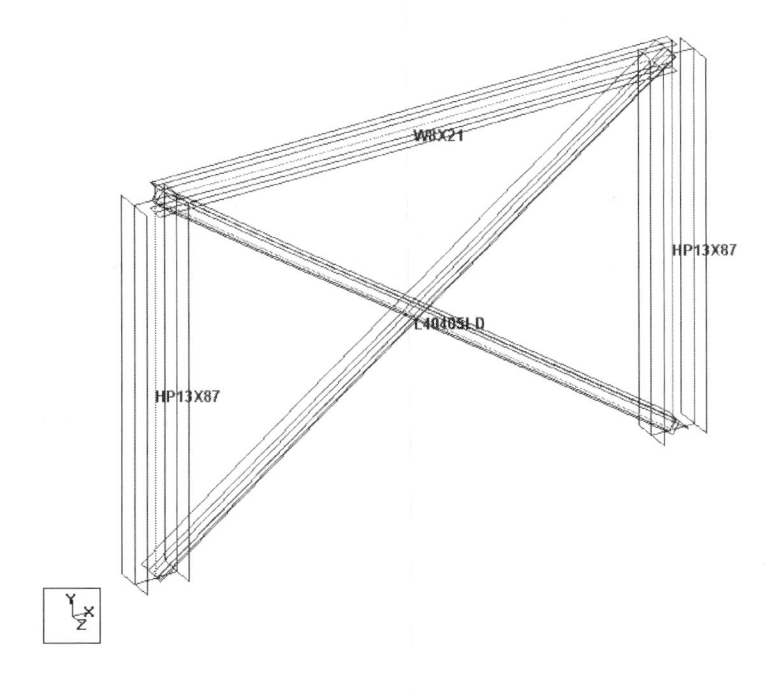

7. Don't save and close.

Type 3: Thickness

- Go to **General** Page Control.
- Make sure you are in the **Property** sub-page.
- Select the desired Plate.
- From Data Area at the right of the screen, click **Thickness**. The following dialog box will appear:

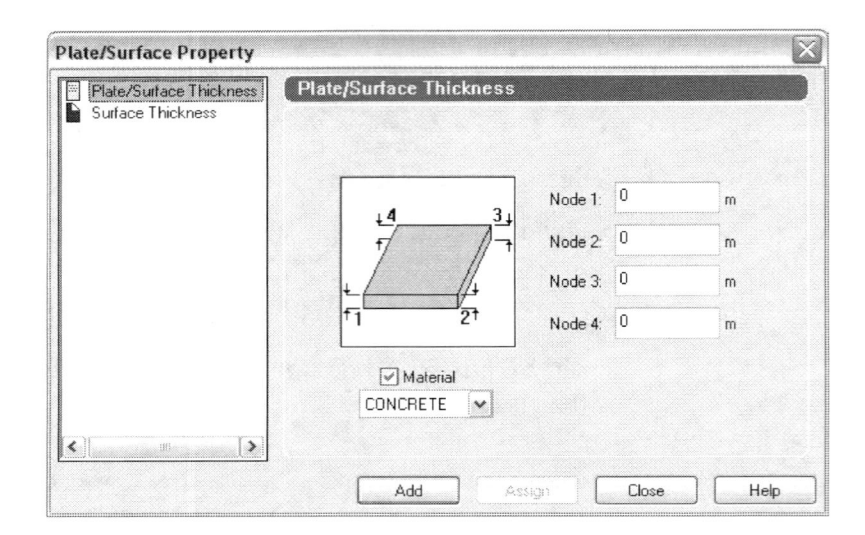

- If you fill Node 1, automatically Node 2, 3, and 4 will be filled with the same value. That applies also for deleting the value of Node 1.
- If you want to give different Thickness for different Nodes, go to each Node, and input the desired value, this will not affect any of the other Nodes.

Note
- Also, you can reach the same command from the menus by selecting **Commands/Plate Thickness**.
- **Material** checkbox is the same as in Prismatic.

General Notes about Property Assigning

- For the three types of Properties mentioned in this Module, there are different functions to help us control the process in a better way:

Double-Click Beam
- After you assign Property for a Beam, double-clicking it will show you a dialog box, select **Property** tab, and the following will appear:

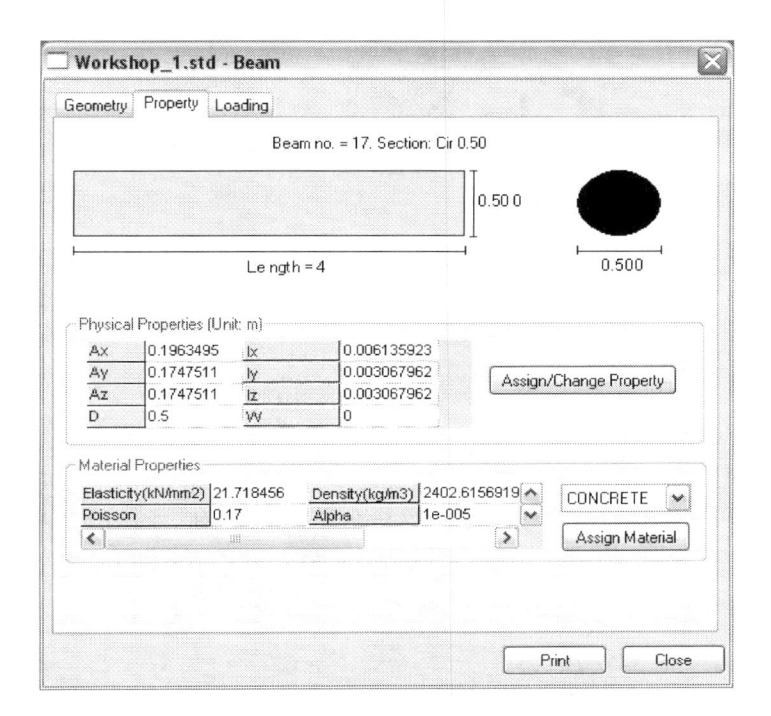

- You will see that STAAD.*Pro* already calculated:
  - Ax, Ay, and Az.
  - Ix, Iy, and Iz.
- If you accepted the built-in Concrete/Steel material constants while you were assigning, you will see Elasticity, Density, Poisson, and Alpha values.

Double-Click Plate ■ The same thing applies if you double-click a Plate after the assignment of Thickness:

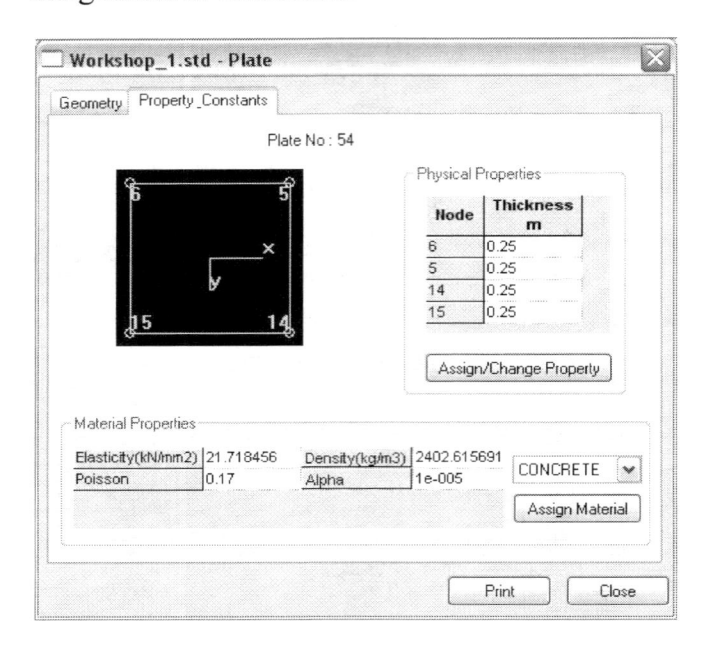

Referencing ■ After you made your assignment, and in the Data Area, you will see the following:

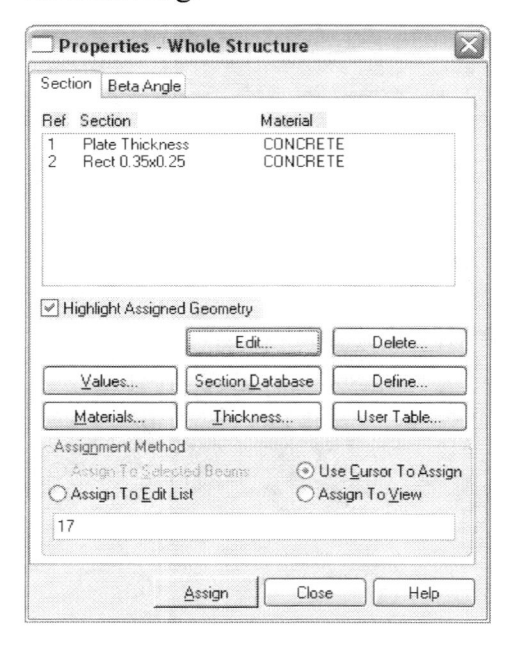

Module 4: Properties

- Each Property assigned will be given a **Ref** number, starting from 1, and it shows if you select the **Material** checkbox or not.
- In STAAD.*Pro* window you will see the name of the section mentioned beside each Beam, or Plate:

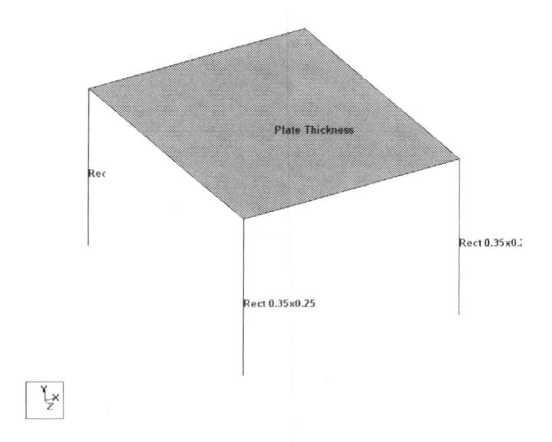

- Clicking on the Ref will automatcially select the Beams, or Plates. In our example if you clicked Ref 2 **Rect0.35X0.25** automatically the columns assigned this Property will be selected, like this:

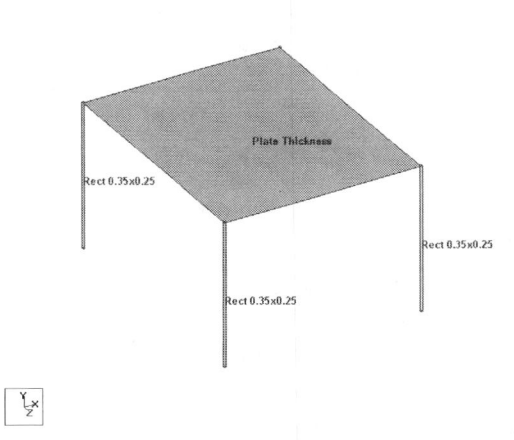

- This is very usefull especially if you want to check your work; that is all the desired Beams are assigned the right cross-section.

Note
- If you don't want the selecting to take place, click off the check box below **Reference** table:

☐ Highlight Assigned Geometry

4-17

Deleting
- If you want to delete an assignement, do one of the following:
  - Select the Ref number you want to delete, and click **Delete** button.
  - Select the Ref number and press **Del** at the keyboard.
- In both way, the following dialog box will appear:

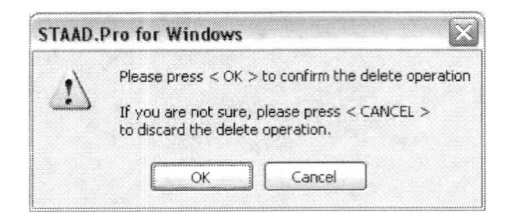

- Click **OK**, to confirm the deletion, or **Cancel** to ignore.

Reference Label
- To change how the Reference Label is displayed on the screen, right-click anywhere in the STAAD.*Pro* window, and select **Labels** from the Properties section,
  - **Reference**, beside each Beam you will see R1, R2, ...etc.
  - **Section**, will show the cross section beside each Beam.
  - **None**, unless you are in the **Property** sub-page, you will not see the Properties Referencing.

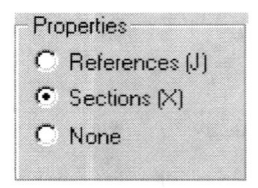

Converting Units By F2
- While you are inputing Prismatic cross section, or Thickness, STAAD.*Pro* allows you to convert dimension from one unit to another, by using F2 key in the edit box.
- Assume your current units is in m, and the avialable dimension is in feet. Instead of converting between the two units, do the following:
  - In the edit box, click F2, the edit box will change to:

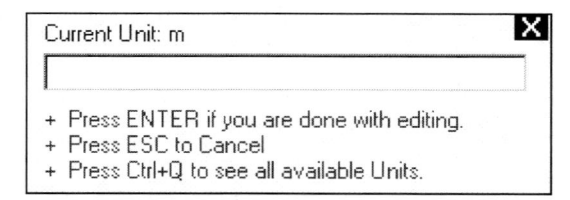

- Type in 1 *ft*, and press ENTER.

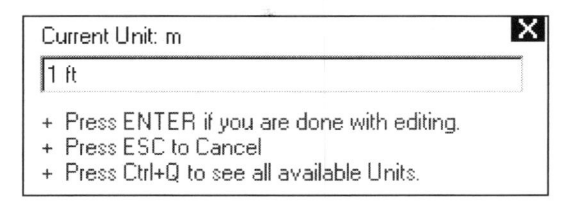

- STAAD.*Pro* will convert it to *m*.

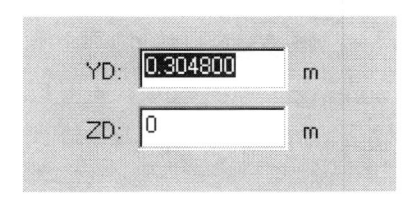

Changing Input Units

- As an alternate method you can also, change your input units, and input the dimension using the new units.

- Assume your current units is in *m*, and the avialable dimension is in *feet*. Instead of converting between the two units, do the following:

  - From the Structure toolbar, select **Input Units**, or from menues select **Tools/Set Current Input Unit**. The following dialog box will appear:

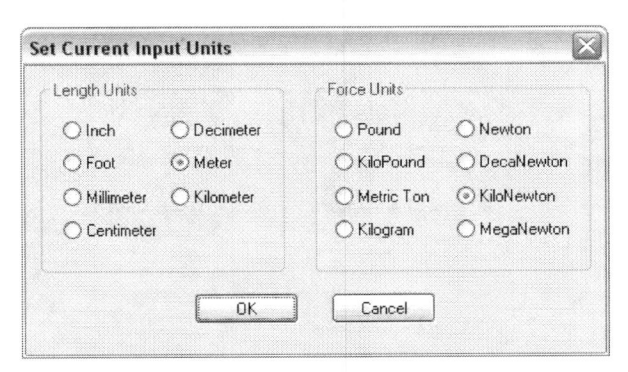

  - Click **Foot**.

  - From now on any dimension will be given to STAAD, will be considerd in *feet*. To get it back to *m*, re-issue the function and change it to *m*.

Assigning Properties to Beams, and Plates (Concrete Structure)

Workshop 2-A

1. Open Small_Building file.
2. **<u>Always select the Material checkbox to be ON.</u>**
3. Select all outer columns, and assign Circular YD=0.50m.
4. Select all inner columns, and assign Rectangular YD=0.5m, and ZD=0.25m.
5. Select all beams, and assign Rectangular YD=0.40m, and ZD=0.20m.
6. Select the Plates, and assign Thickness=0.135m
7. This is the rendered image of the structure:

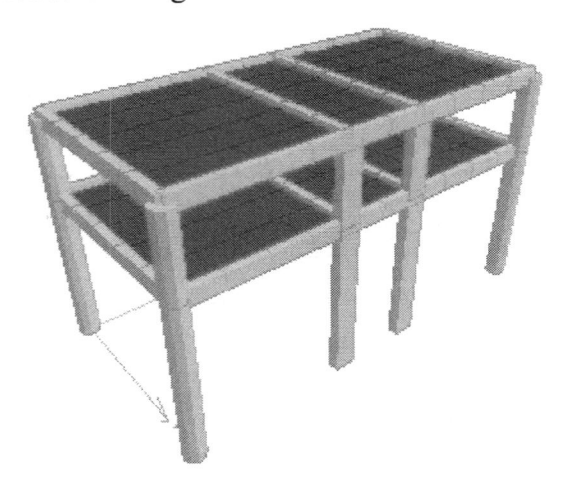

8. Save and Close.

Assigning Properties to Beams, and Plates (Steel Structure)

Workshop 2-B

1. Open Small_Building file.
2. Using **Insert Node** function cut the Beam as indicated below to three Beams with each 1m long:

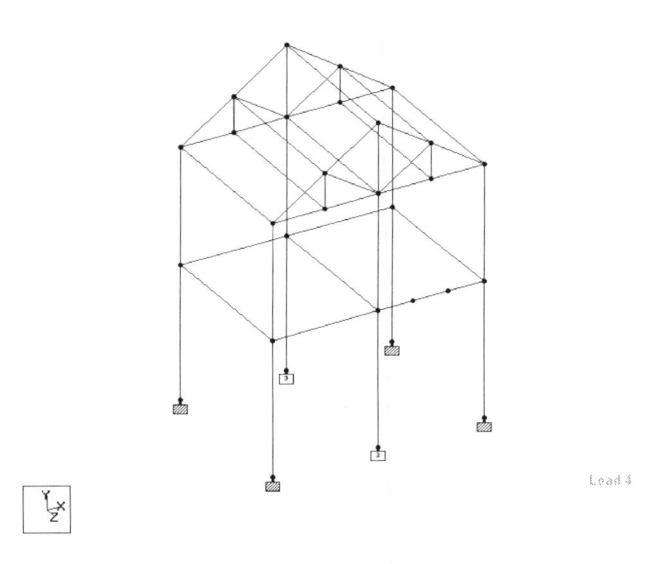

3. **<u>Always select the Material checkbox to be ON.</u>**
4. Assign the following cross-section:
    a. For top chord (left and right, front and back) assign W10X15

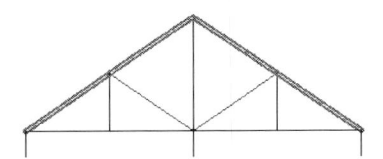

4-21

b. For Truss Members (front and back) (see the illustration), assign L20205

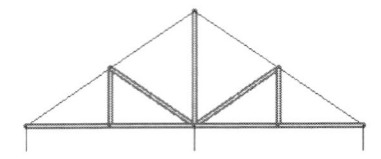

c. For Connections (see the illustration) assign S6X12

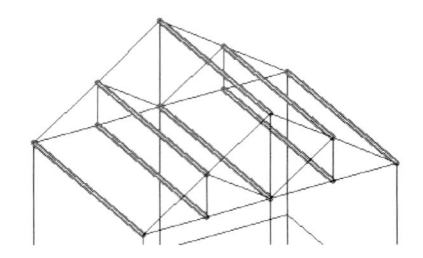

5. For all columns assign HP14X102
6. For all beams assign W16X36
7. This is the rendered image of the structure:

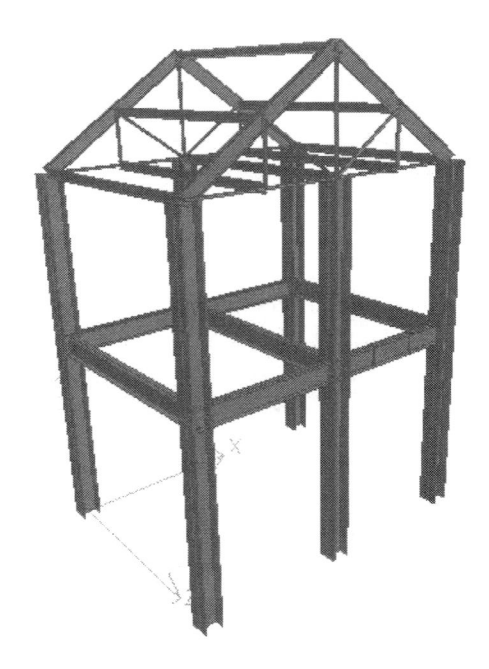

8. Save and Close.

Module Review

1. You can't assign a Prismatic cross-section to Steel Beam:
    a. True
    b. False

2. Instead of inputting cross section I can input Ix, Iy, Iz:
    a. True but only for circular cross-section.
    b. True for all sections.
    c. False you have to input cross-section always.
    d. Any of the above.

3. Using **Structure Diagrams**, and under _____ you can view the Section Outline of your cross-section.

4. While you are inputting Prismatic cross-section you can convert to any unit you want:
    a. True
    b. False

5. For each section in a Steel table you should specify:
    a. Cover Plate thickness.
    b. Proper Specification depends on the category belongs to.
    c. Space between two Angles
    d. All of the above.

6. For Plates, specify the _____ as property.

Module Review Answers

1. a
2. b
3. 3D Sections
4. a
5. b
6. Thickness

Module 5:

Constants, Supports, and Specifications

This module contains:

- Material Constants
- Geometric Constant
- Supports
- Beam, and Plate Release
- Truss Members

Introduction

- There are two types of Constants in STAAD.*Pro*:
    - Material Constants.
    - Geometric Constants.
- Supports, is an essential part of the input file. STAAD.*Pro* offers several types of Supports.
- You can set different Specifications for Beams, and Plates. This will help STAAD.*Pro* to better understand your Model and hence produce the correct results.

Material Constants

- In Module 4, we found a checkbox in all of the **Property** dialog boxes to include a certain Material (Concrete, or Steel) Constants with the cross-section.
- If this checkbox was checked ON, this means the user is accepting the default values in STAAD.*Pro*:

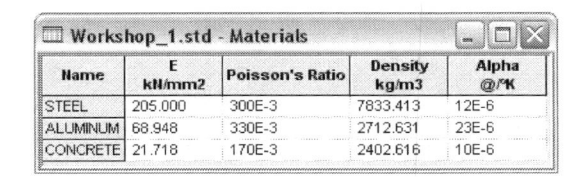

Name	E kN/mm2	Poisson's Ratio	Density kg/m3	Alpha @/°K
STEEL	205.000	300E-3	7833.413	12E-6
ALUMINUM	68.948	330E-3	2712.631	23E-6
CONCRETE	21.718	170E-3	2402.616	10E-6

- If this checkbox was checked OFF, user should input the Material Constants manually. At least two are needed:
    - **E**, Young's Modulus of Elasticity (always required).
    - **Density**, if you want STAAD.*Pro* to calculate selfweight.
- You can ignore **Poisson's Ratio**, and **Alpha** (Thermal Expansion factor—for Thermal loading only).
- There are three ways to assign Material Constants to Beams, or Plates:
    - Using menus
    - Using Page Control
    - Using double-click

Using menus
- Select desired Beams, or Plates.
- From menus select **Commands/Material Constants**, then select one of the following choices:
  - Density
  - Elasticity
  - Poisson's Ratio
  - Alpha
  - Damping Ratio
  - G (Shear Modulus)
- Regardless of the constant you want to input, you will get almost the same dialog box:

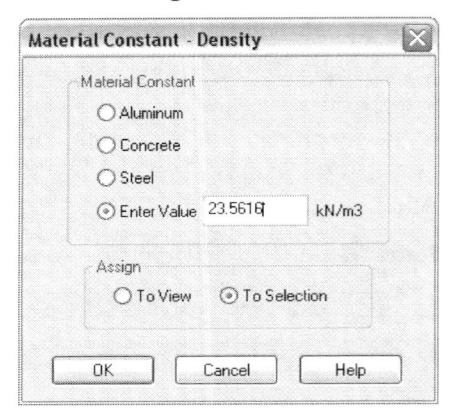

![Material Constant - Density dialog box showing options for Aluminum, Concrete, Steel, or Enter Value 23.5616 kN/m3, with Assign options To View or To Selection, and OK, Cancel, Help buttons]

Note
- If you selected Beams, or Plates before the command, then under **Assign**, select **To Selection**.
- You can assign to all Beams, and Plates, by selecting To View.
- While you are in of the following dialog boxes, you can use F2 key to convert units (just as we learned in the Module 4):
  - Density
  - Elasticity
  - G (Shear Modulus).

Module 5: Constants, Supports, and Specifications

Using Page Control
- From **General** Page Control, select **Material** sub-page.
- At the Data Area, click **Isotropic** tab.

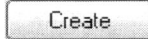

[Create]
- Click **Create** button, to create a new material constants, you will get the following dialog box:

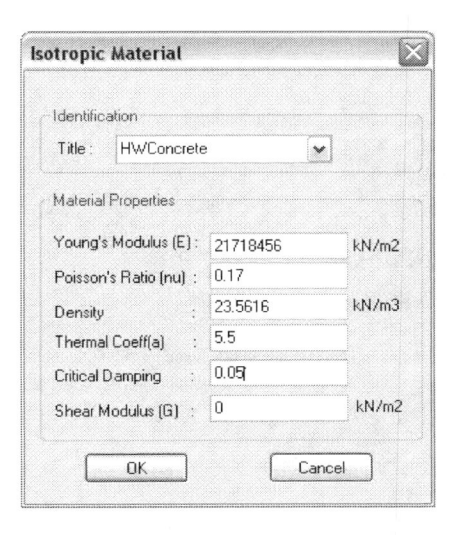

Isotropic Material

Identification
Title: HWConcrete

Material Properties
Young's Modulus (E): 21718456 kN/m2
Poisson's Ratio (nu): 0.17
Density: 23.5616 kN/m3
Thermal Coeff(a): 5.5
Critical Damping: 0.05
Shear Modulus (G): 0 kN/m2

[OK] [Cancel]

- Type the name of this material for future usage (in this file only), input the data you want, and click **OK**, a new Material will be added.
- While the new material is selected, select desired Beams, or Plates, and click **Assign**.

Using double-click
- Double-click on the desired Beam, Plate, a dialog box will appear
  - Select **Property** tab.
  - Under **Material Property**, there will be a pop-up list.
  - Select the desired Material.
  - Click Assign Material.

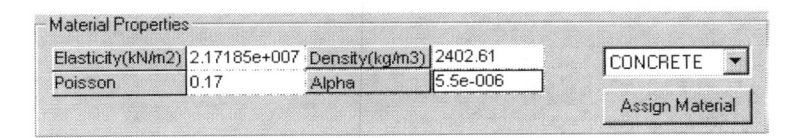

Material Properties
| Elasticity(kN/m2) | 2.17185e+007 | Density(kg/m3) | 2402.61 |
| Poisson | 0.17 | Alpha | 5.5e-006 |

CONCRETE
[Assign Material]

Note
- If you already created an Isotropic Material, you will find it in the list.

5-5

Assign Material Constants

Exercise 20

1. Open Common.std.
2. Select all the structure and assign YD=0.6m, and ZD=0.2. Make sure to check **Material** checkbox OFF.
3. Select the three Beams.
4. From menus assign the following values for all Beams:
   a. E=21 KN/mm^2
   b. Density=23.534X10^{-9} KN/mm^3

Note
- We used mm and not m, so you need to make a conversion, so change the Input Units to mm.
- In order to input a vlaue with exponential like 23.534X10^{-9} use letter **E**, like 23.534E-9 (also you can use **e**)
5. Don't close the file and keep it open for the next exercise.

Geometry Constant

- When you draw a 3D frame, then you are dealing with two main 2D frames.
- Accordingly when you assign a cross-section to any Beam (beam, or column) STAAD.*Pro* will orient the cross-section in a way that it may not satisfy your structural case.

Example
- Assume the following 3D frame:

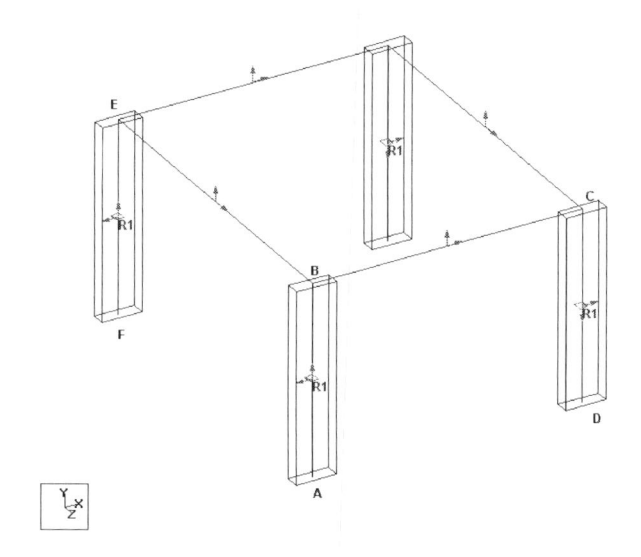

- Note the orientation of the cross-section, and hence the orientation of local Y.
- This assumption works perfect for the loading to be working at frame ABCD, and not ABEF.
- Let's assume that the loading system you have doesn't work at ABCD frame, but instead at ABEF.
- Your requirement now will be to rotate the cross-section in a way that the higher moment of inertia will be at ABEF.
- Here when it comes Beta (β) role.
- When $\beta = 90°$ this means you will rotate the cross-section by 90°.

- It will work with Concrete and Steel cross-sections.
- The frame will look like the following:

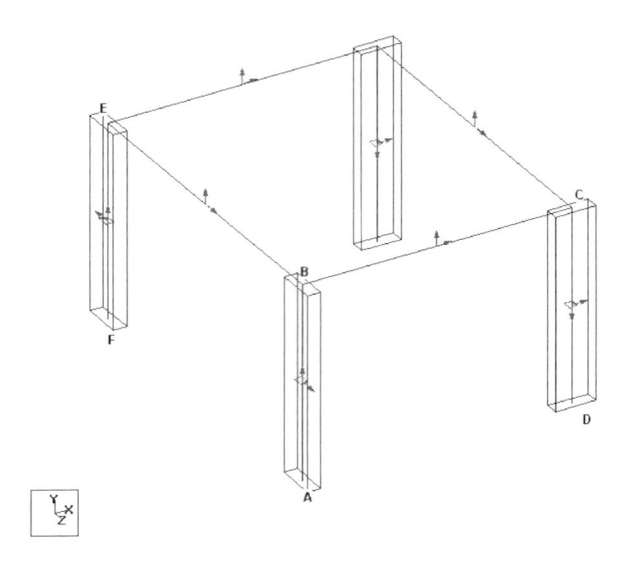

- The cross-section of AB was rotated +90°.
- The cross-section of EF was rotated -90°.
- This is because we want local Y to point outside.

Note
- User can use any angle, β is not only 90°, or -90°.
- Example on using -45°:

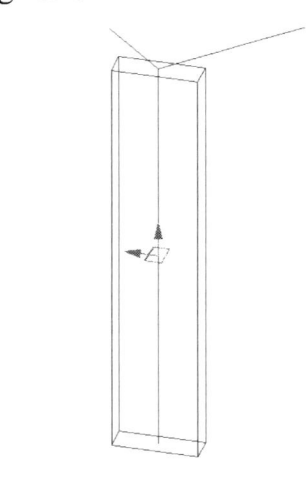

- Angles in STAAD.*Pro* is Absolute, and not Relative, meaning 45° angle is always 45° and not relative to the current position of the Beam.

Module 5: Constants, Supports, and Specifications

- There are three ways to assign β to Beams:
  - Using menus.
  - Using Page Control.
  - Using double-click.

Using menus
- Select desired Beams.
- From menus select **Commands/Geometric Constants /Beta Angle**, the following dialog box will appear:

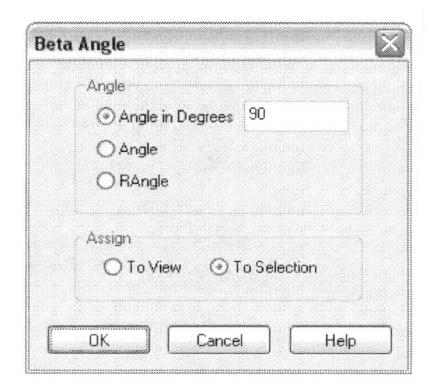

- Input the desired angle (Make sure **To Selection** is selected).

Using Page Control
- Select th desired Beams.
- From **General** Page Control, **Property** sub-page, from Data Area, click on the **Beta Angle** tab.
- Data Area will change to:

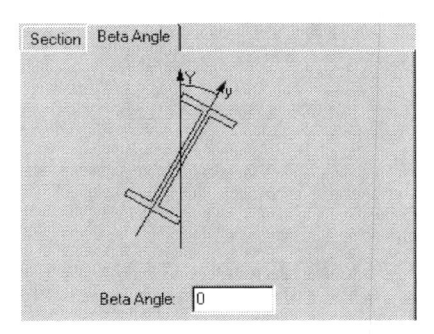

- Input the desired angle, and click **Assign**.

5-9

Using double-click
- Double-click on the desired Beam. A dialog box will appear, under the **Additional Info** part, the current Beta Angle will be displayed.

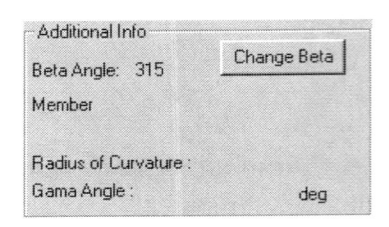

- Click **Change Beta** button, and set the new Beta Angle.

Assign Geometric Constant

Exercise 21

1. Continue with the previous file.
2. Using **Structure Diagrams**, show the **Section Outline**.
3. Select the beam (the horizontal member).
4. Set **Beta** angle to be 90.
5. Don't close the file and keep it open for the next exercise

Module 5: Constants, Supports, and Specifications

Supports

- STAAD.*Pro* provides several types of supports to model any structural case.
- Supports are essential part of the input file. Without this part, the Analysis may not run, and may produce an error message.

Types of Supports
- We will focus in this tutorial on the following supports:
  - Fixed
  - Pinned
  - Fixed But

Fixed Support
- Fixed in STAAD.*Pro* means there will be no movement in any direction, and no rotation around any axes.
- There will be six reactions on this support: F_x, F_y, F_z, M_x, M_y, and M_z.
- Used mostly to model isolated footings.
- Can be used with STAAD.etc to design the footing after the analysis results generated from STAAD.*Pro*.

Pinned Support
- Pinned in STAAD.*Pro* means there will be no movement in any direction, but there will be rotation around all axes.
- There will be three reactions only on this support: F_x, F_y, F_z.
- Used mostly in Plane structure (2D geometry parallel to XY Plane). If used in Plane structure, there will be only 2 reactions F_x, and F_y (which is Axial, and Shear respectively) as F_z will not be considered by the structure it self. If you want to use it in 3D geometry, study your case carefully, as it may not fit what you need.

Fixed But Support
- We recommend this support to be used in cases of releasing your fixed support, especially in 3D frames.
- User can control which of the three forces, or the three moments to be released.
- This will give the user the power to model the structural case exactly.
- You can release any of the following: F_x, F_y, F_z, M_x, M_y, M_z

How to Assign Supports

- There are two ways to assign Supports to Nodes:
  - Using menus.
  - Using Page Control.

Using menus
- Select desired Nodes.
- From menus select **Commands/Support Specifications**, then select one of the first three choices; **Pinned, Fixed, Fixed But/Spring**.
- For **Fixed** and **Pinned**, there will no additional information, hence click **Assign**.
- For Fixed But, you should select which of the 3 forces, and/or which of the 3 moments to be released:

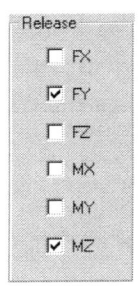

```
Release
  ☐ FX
  ☑ FY
  ☐ FZ
  ☐ MX
  ☐ MY
  ☑ MZ
```

Using Page Control
- From **General** Page Control, select **Support** sub-page.
- Once you are there, the cursor will change to Node cursor automatically.
- Select the desired Nodes.

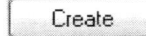

[Create]
- From the Data Area, click **Create**.
- A dialog box will be displayed, pick the desired type of supports, and click **Assign**.

Note
- The dialog box in the two methods is the same.
- In the case of **Fixed But**, it will be meaningless if you clicked all the 3 forces and the 3 moments to be released.

Editing Supports

- From the **Selection** toolbar, select **Support Edit Cursor** icon, double-click the support, to show the dialog box of the support to edit it.
- Another way would be to double click the Support definition in the Data Area, to show the dialog box of the Support to edit it.

Deleting
- If you want to delete an assignment, do one of the following: 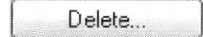
    - Select the Ref number of the support you want to delete, and click Delete button.
    - Select the Ref number of the support and press **Del** at the keyboard.
- In both way, the following dialog box will appear:

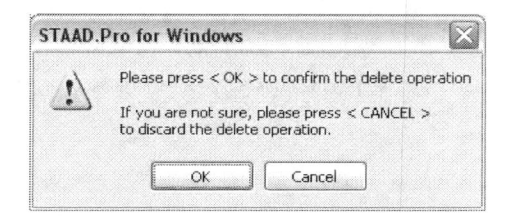

- Click **OK**, to confirm the deletion, or **Cancel** to ignore.

Assign Supports

Exercise 22

1. Continue with the previous file.
2. Go to **General** Page Control.
3. Select **Support** sub-page.
4. Select the left support, and define it as **Fixed**.
5. Select the right support, and define it as **Pinned**.
6. Don't close the file and keep it open for the next exercise.

Specifications

- **Specifications** command in STAAD.*Pro* is the way to define things about your Model that geometry alone can't.
- For example, by default all of your Nodes at the ends of each Beam are stiffed. In **Specifications** you can release Beams from any side using any of the six reactions.
- Another example would be to declare to STAAD.*Pro* that some of your Beams in Space model are Truss members; hence they can sustain axial loads only.

Beam Release
- The Nodes at the ends of each Beam, is always considered to be rigid Nodes, hence there will be six reactions at each Node.
- You can release a Beam by releasing the **Start** Node, or the **End** Node from any of the 3 forces, and/or 3 moments.
- Point to the desired Beam, the Beam number will be displayed along with the Green color (Start), and Blue color (End). This way you will know where exactly you will define the release of the Beam.
- Select **General** Page Control, and **Spec** sub-page.

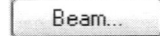

- Select the desired Beam, then click **Beam** button.
- You can reach the same command using menus by selecting **Command/Member Specification/Release**.
- The following dialog box will appear:

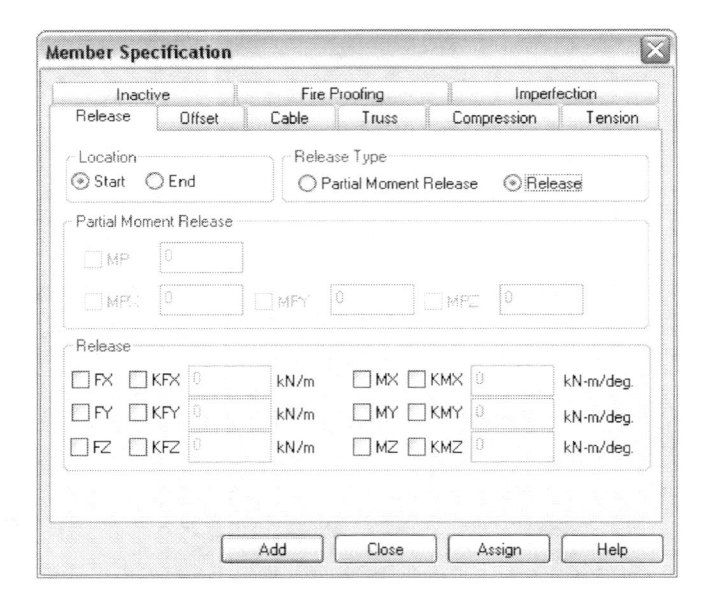

5-14

Module 5: Constants, Supports, and Specifications

- Specify the following information:
  - The Release Type, whether Partial Moment Release, or Release.
  - The **Location**, whether at the **Start**, or at the **End**.
  - Specify the **Release**, choose the desired checkbox.
  - Click **Assign**.
- This is how STAAD.*Pro* represent the Release:

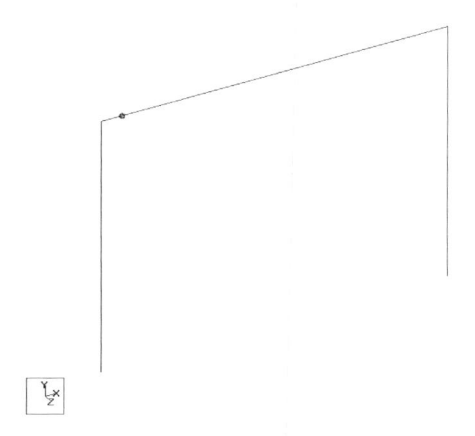

Editing Beam Release

- After you apply the release you can change it, by double clicking the Beam, a dialog box will appear, under **Releases** you will see the current release conditions:

- Click **Change Releases At Start** button, if the current release is at the Start, or else click the other key.
- The original Release dialog box will appear, and you can change it to the desired release.
- From the **Selection** toolbar, select **Member Release Edit** Cursor icon, double-click the Release itself (and not the Beam), to show the dialog box of the Release to edit it.
- Another way would be to double-click on the Release from the Data Area, to show the dialog box and edit the Release.

Plate Release
- The same idea applies for Plates. In this case you can specify 3-4 Nodes instead of 2 only. Select **General** Page Control, and **Spec** sub-page.

[Plate...]
- Select the desired Plate, then click **Plate** button.
- You can reach the same command using menus by selecting **Command/Plate Element Specification /Release**.
- The following dialog box will appear:

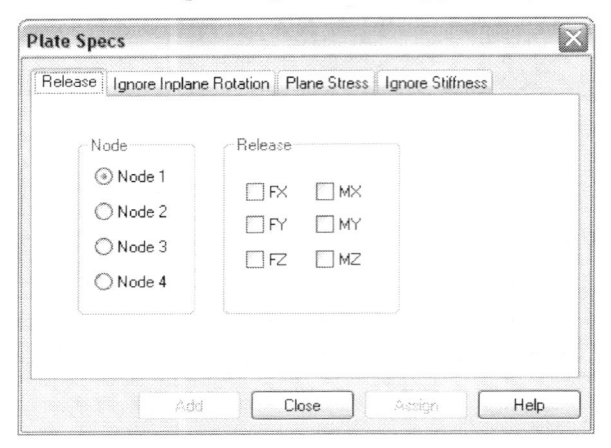

- Select the Node you want to release (they numbered 1, 2, 3, and 4)
- Then specify the Release by clicking the checkbox on. Click **Assign**.
- If you want to release another Node in the same Plate, without closing the dialog box, select the other Node, and set the new release, and click **Assign**.
- To edit it the release you can double-click the Plate, and follow the same procedure as in the Beam case.

Deleting
- If you want to delete an assignment, do one of the following:

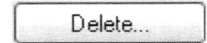

[Delete...]
  - Select from the list the Specification you want to delete, and click **Delete** button.
- Select the Ref number of the support and press **Del** at the keyboard. In both way, the following dialog box will appear:

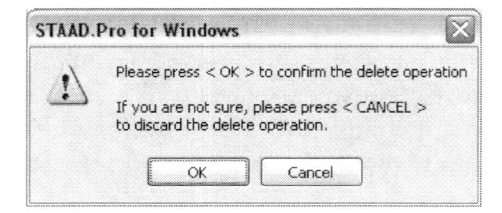

- Click **OK** to confirm the deletion, or **Cancel** to ignore.

Module 5: Constants, Supports, and Specifications

Truss Members
- If user defined the file as **Truss** structure, all the Beams (by default) will take only axial loads only. Hence the Beams, will not carry neither shear nor moment.
- But if the user defined the structure as **Space**, or **Plane**, all the Beams (by default) will carry axial, shear, and moment.
- Accordingly, if you have in **Space**, or **Plane** structure a truss, STAAD.*Pro* will never know how to deal with truss members, except to consider them as any other members, which means will carry axial, shear, and moment.
- To let STAAD.*Pro* treat truss members as special case, you have to specify that as **Specification**.
- From **General** Page Control, select **Spec** sub-page.
- You can reach the same command using menus by selecting **Command/Member Specification/Truss**.
- Select the Beams to be declared as Truss members. 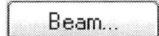
- Click **Beam** button. The **Member Specification** dialog box will appear, select **Truss** tab.

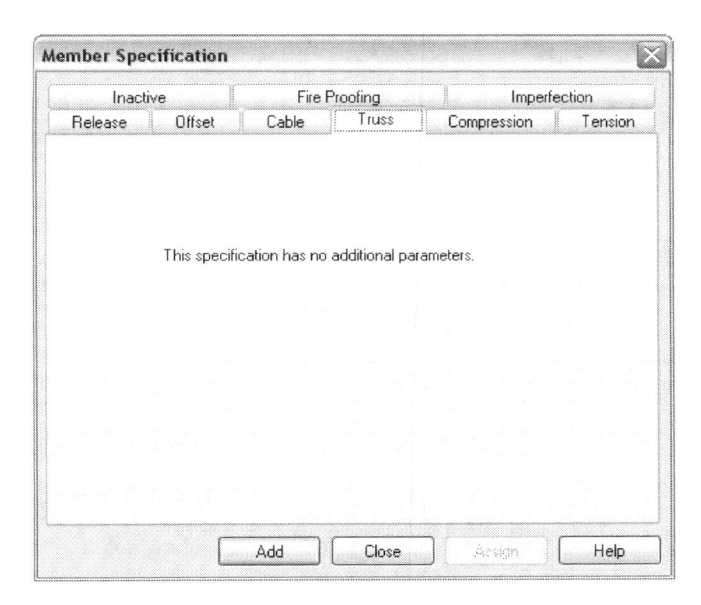

- Click **Assign**.

5-17

- This is how STAAD.*Pro* represent the Truss members:

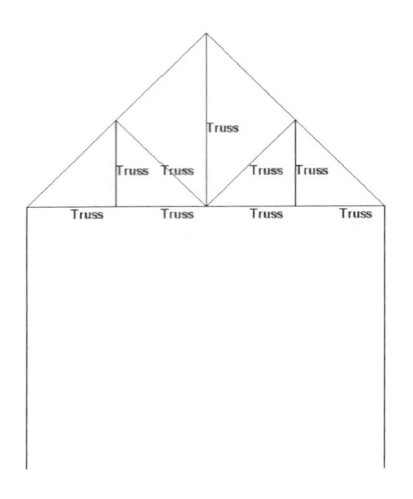

Assign Specifications

Exercise 23

1. Continue with the previous file.
2. Go to **General** Page Control, and select **Spec** sub-page.
3. Point to the beam (the horizontal member). Make sure you knew the Start and the End of it.
4. Select it, and click **Beam**.
5. Release it from Mz only from the End.
6. Click **Assign**.
7. Don't save and close.

Assigning Constants, Supports, and Specs (Concrete Structure)

Workshop 3-A

1. Open Small_Building file.

2. Right-click, and select from the shortcut menu, **Structure Diagrams**, when the dialog box appears select from **3D Section**, the option **Section Outline**. We did this step so we can see the cross section before and after the Beta angle assignment.

3. Select the Beams as shown:

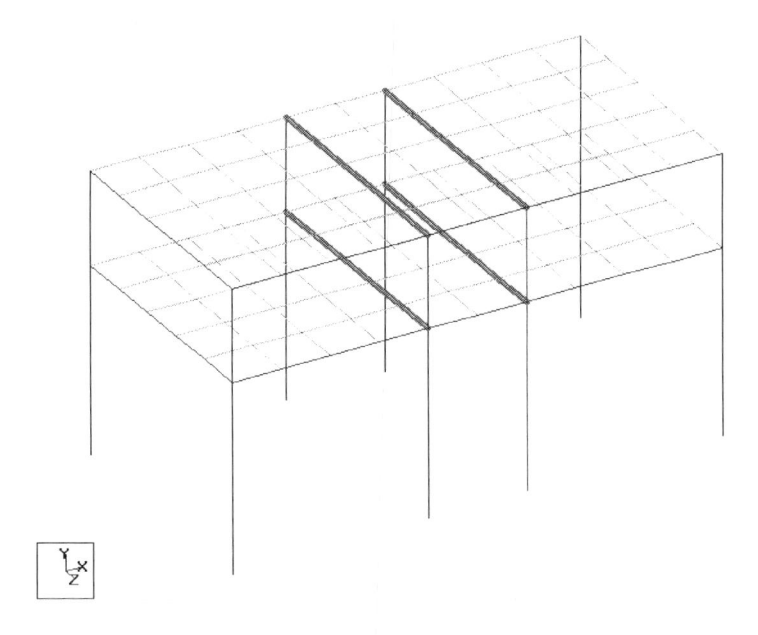

4. Set the Beta Angle to be 90.

5. Define all supports as Fixed.

6. Save and Close.

Assigning Constants, Supports, and Specs (Steel Structure)

Workshop 3-B

1. Open Small_Building file.
2. Select the Beams as shown:

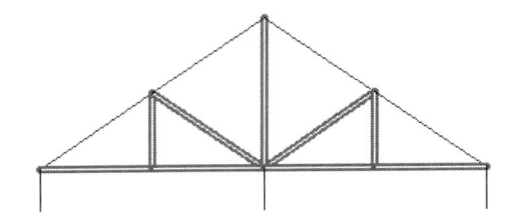

3. Declare the selected Beams as Truss Members.
4. Define outer supports as Fixed, and the inner supports as Fixed but Mz.
5. Save and Close.

Module Review

1. Beta must be always 90°:
    a. True
    b. False

2. STAAD.*Pro*:
    a. Has a facility to declare Truss members.
    b. In Space model can't differentiate between Beams, and Truss Members.
    c. Will treat Truss members as ordinary members in Space structure.
    d. All of the above.

3. Pinned support will have _____ reactions.

4. You can release a Beam from any of the 6 reactions at the Start or End.
    a. True
    b. False

5. For Plates, one of the following is true:
    a. In order to release a Plate, you can release the Beams holding the Plate.
    b. You can release the Nodes of the Plates.
    c. You can release more than one Node of the Plates.
    d. Answers b & c.

6. _____, and _____ should be input as Material constants the rest can be ignored.

Module Review Answers

1. b
2. d
3. 3
4. a
5. d
6. E & Density

Module 6:

Loading

This module contains:

- Introduction to the types of loading in STAAD.*Pro*
- How to create Primary Loads
- Individual Loads: Member Loads
- Individual Loads: Area & Floor Loads
- Individual Loads: Plate Loads
- Individual Loads: Node Loads
- How to View and to Edit Loads
- How to create Combinations (Manual or Automatic)

Introduction

- Loading is considered to be the last step in creating your input file before the Analysis Command.
- Loading will be done through creation of Primary Loads, which includes Individual Loads.
- Then creating Combination Loads, which will combine a factored Primary Loads, simulating the design codes combinations, which will generate the maximum shear/moments results.

Primary Loads
- Primary Load is the base for the loading in STAAD.*Pro*.
- Each Primary Load should have a **Number** (essential to STAAD.*Pro*), and a **Title** (optional for STAAD.*Pro* but important to the user).
- As an example of Loading Number and Title will be: Load 1 Dead Load, Load 2 Live Load, and Load 3 Wind Load.
- If you left **Title** empty, STAAD.*Pro* will accept that, but later on, it will be very difficult, to remember what is this load, hence make sure always to type a good title.
- Primary Load contains inside it all the individual loads which may act on:
  - Nodes
  - Beams
  - Plates
- One of the pre-defined loads, which can be included in a Primary load, is **Selfweight**.

Combination Loads
- Combination Load contains Primary Loads defined earlier multiplied by a factor.
- The number of the first Combination will take the following number of the last Primary Load. As an example, if the last Primary Load holds number 3, the first Combination will hold number 4.
- One of the Combination may look like:
    - 1.4 * (1) + 1.6 * (2)
- Because STAAD.*Pro* deals with the number of the Primary Loads, and not the title, therefore the equation of the Combination should show only the numbers of the Primary Loads.

How to Create Primary Loads

From Page Control
- Go to **General** Page Control, and select **Load** sub-page, the following dialog box will appear on the Data Area:

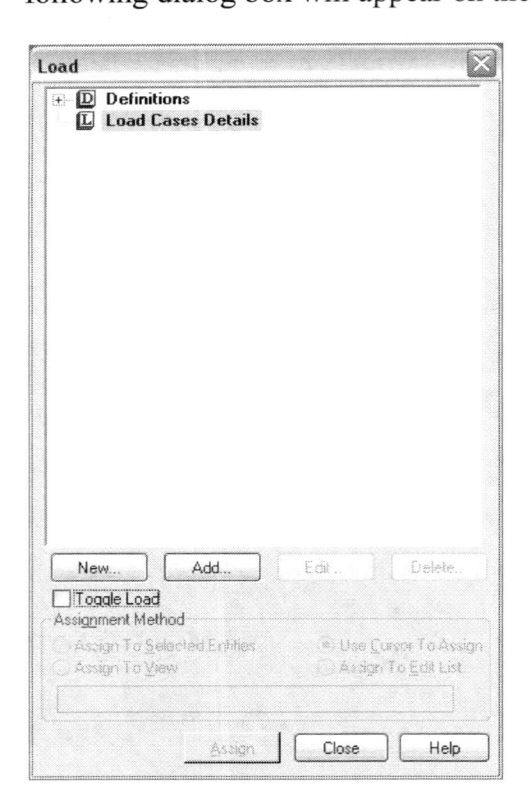

Module 6: Loading

[Add...] ■ To add a new Primary load, click **Add** button, the following dialog box will appear:

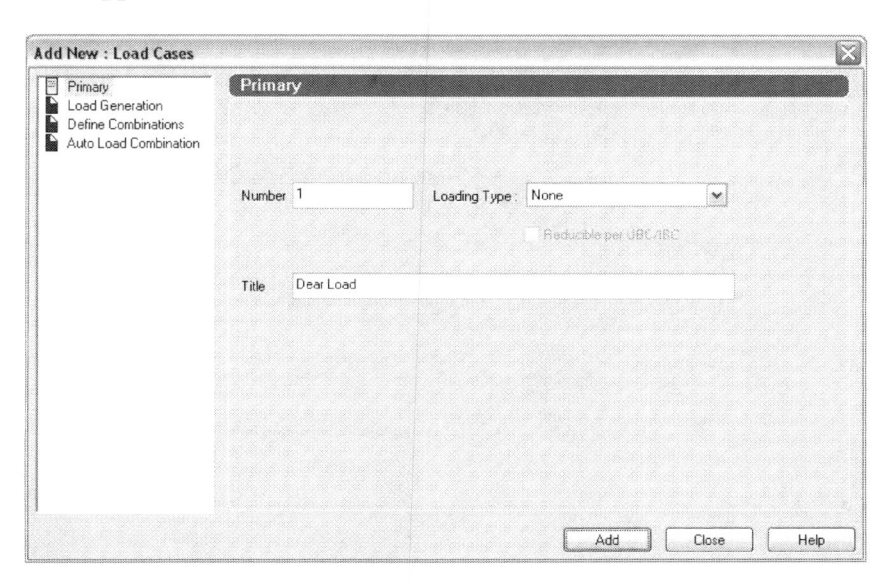

■ The **Number** of the Primary load is given automatically.
■ Leave the **Loading Type** to **None** (we will discuss this later)
■ Type in a good **Title**.
■ Click **Add** button.
■ A new load case will be added, see the figure below:

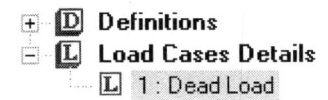

```
[+] D Definitions
[-] L Load Cases Details
      L 1 : Dead Load
```

From Menus ■ From the menus, select **Commands/Loading/Primary Load**, wherever you are, this command will take you directly to **Load** sub-page, and open the dialog box shown in the above to define/edit Primary Loads.

6-5

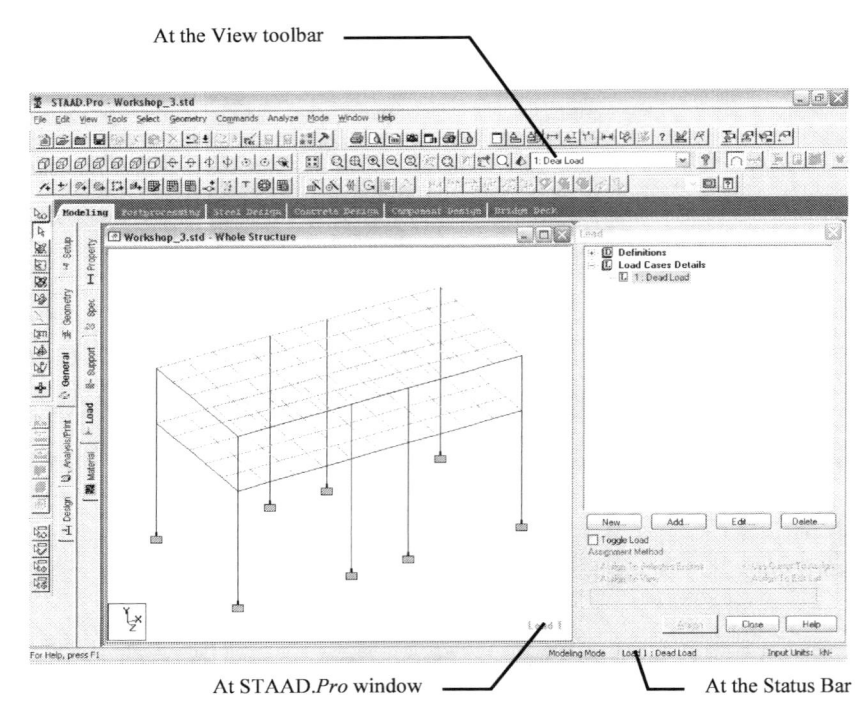

Note ■ You can see the number of the current load in 3 different places:

At the View toolbar

At STAAD.*Pro* window — At the Status Bar

Individual Loads: Introduction

- While you are in a Primary Load, you will able to apply loads to Nodes, Beams, and Plates.
- It is preferable to select the desired Nodes, Beams, or Plates prior to applying the load. This way you will be able to apply the load to mulitple Nodes, Beams, or Plates which will minimize your time.
- STAAD.*Pro* enables the user to specify the direction of each load using Global, Local, or Projected methods.
- STAAD.*Pro* is equipped with a **Selfwheight** command, which can calcualte selfweight of the strcuture based on cross-section, length, and material density.
- STAAD.*Pro* can simulate a One-Way and Two-Way slab loading directly on the beams carrying any slab.

Individual Loads: Selfweight

- No need to select anything, neither Members, nor Plates.
- Select the desired primary load, and click **Add** button, the following dialog box will appear:

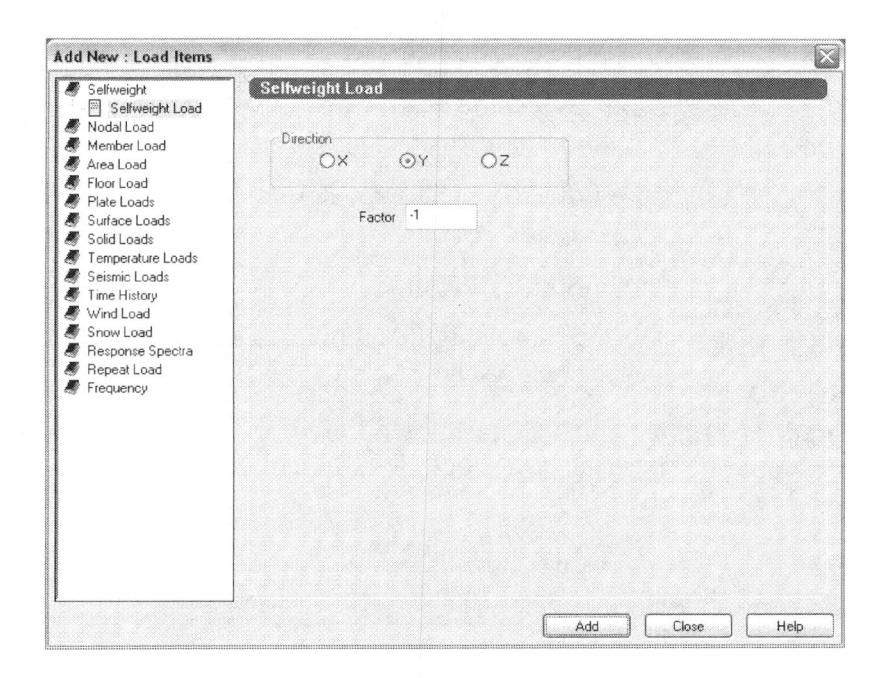

- By default the **Y** is the **Direction** of the load, and the **Factor** is **-1**.
- Click **Assign**.

Individual Loads: Member Loads

5-steps procedure
- Member Loads are loads applied to Beams, covering all types, like, beams, columns, truss members, bracing members, etc.
- It is preferable to select the Beams prior to invoking the command.
- This is a very simple approach to applying load to Beams:
  - Select the Beams.
  - Select the desired primary load case, and click **Add** button, a dialog box will appear.
  - Specify the type, value of the load, and the distances (if applicable).
  - Specify the direction of the load, click **Add** button.
  - From Assignment method select **Assign to Selected Beams**, then click **Assign** button.

Concentrated Force or Moment
- Click Concentrated Force, or Concentrated Moment.
- The case should look like the following:

- The following dialog box will appear:

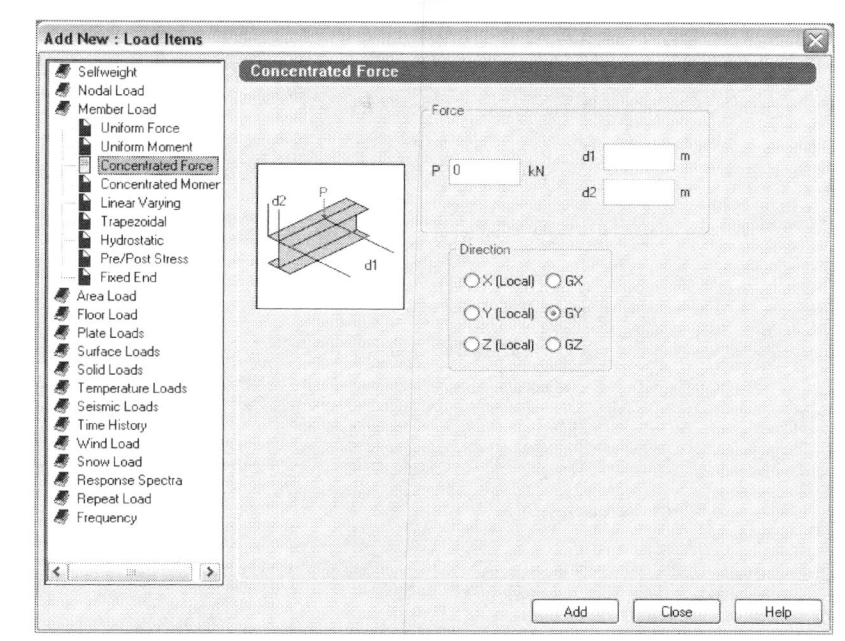

- Specify **P** force value, **d1**, and **d2**:
  - The sign of the force will be discussed in the **Force Direction** part. **d1** as shown in the illustration is measured from the **Start** of the Beam. **d2** is the **eccentricty**.
  - If d1, and d2 left blank, the load is at the middle of the Beam.

Uniform Force or Moment
- Click Uniform Force, or Uniform Moment.
- The case should look like the following:

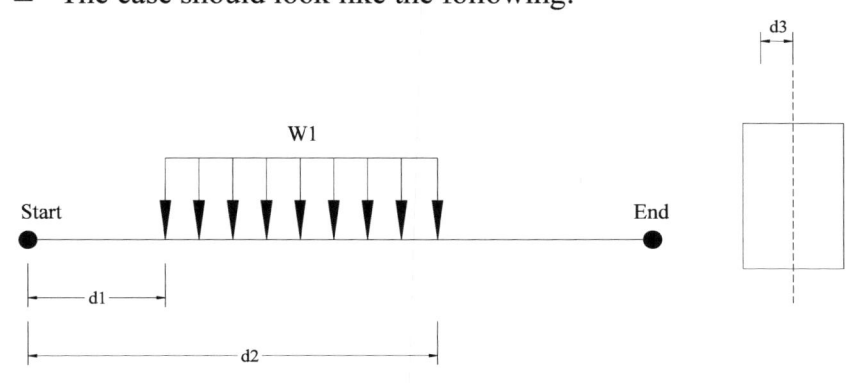

- The following dialog box will appear:

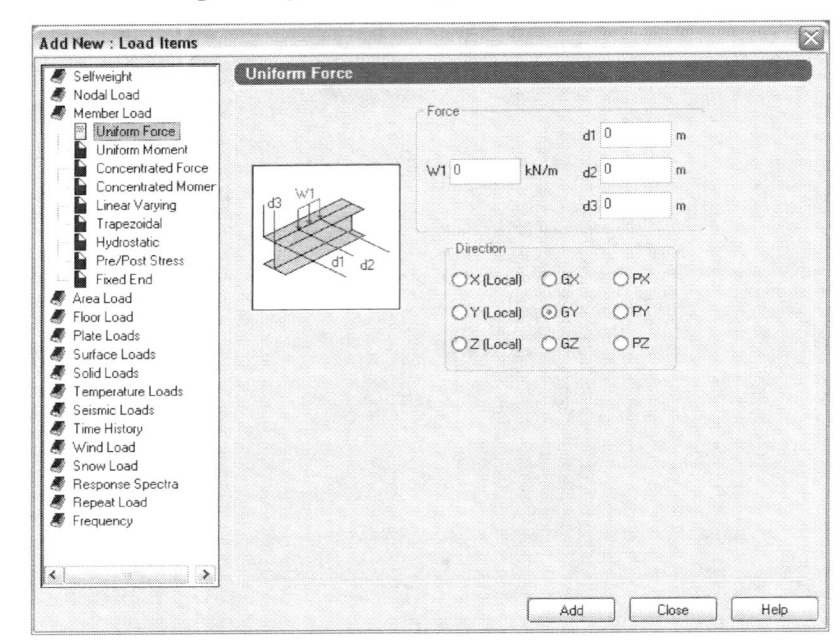

- Specify **W1** force value, **d1**, **d2**, and **d3**:
  - The sign of the force will be discussed in the **Force Direction** part. **d1** and **d2** as shown in the illustration is measured from the **Start** of the Beam. **d3** is the **eccentricty**.
  - If d1, d2, and d3 left blank, the load will cover the whole Beam.

Trapezoidal Force
- Click Trapezoidal tab.
- The case should look like the following:

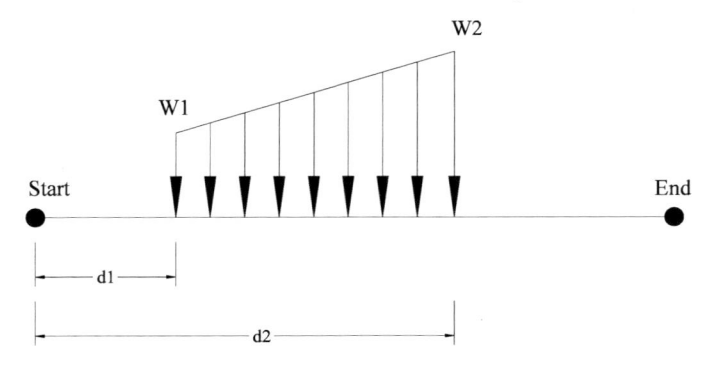

6-10

Module 6: Loading

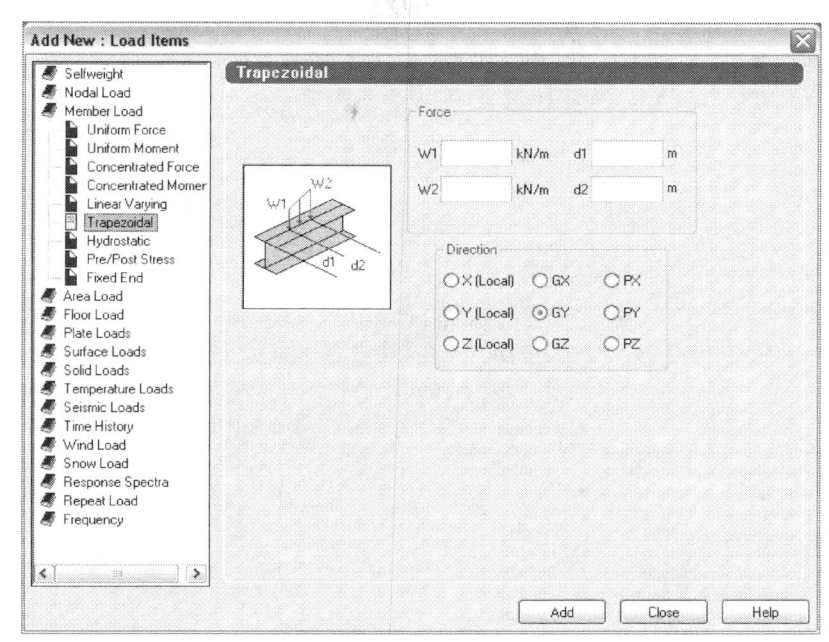

- The following dialog box will appear:

- Specify **W1**, and **W2** force value, **d1**, and **d2**:
  - The sign of the force will be discussed in the **Force Direction** part. **d1** and **d2** as shown in the illustration is measured from the **Start** of the Beam. no **eccentricty** in this type of loading.
  - If d1, and d2 left blank, the load will cover the whole Beam.

Force Direction
- In the above three types of Individual loads, 9 options (except for Concentrated load 6 were available only) to specify the direction of the load:

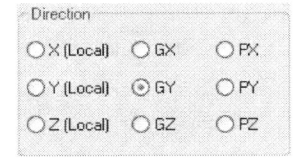

- Three of them are in the **Local** direction of the Beam, X, Y, and Z.
- Three of them are in the **Global** direction, GX, GY, and GZ.
- Three of them are in the **Projected** direction, PX, PY, and PZ.
- Specify whether the desired load will act parallel to the Globals, or Locals, or in Projected fashion. According to your decision, specify whether it holds negative sign or positive sign. Here are some examples:

6-11

Examples for Force Direction

- In case of a beam as in the illustration below, note that; if you specify the direction of the load as GY or Y, the sign of the load will be negative.

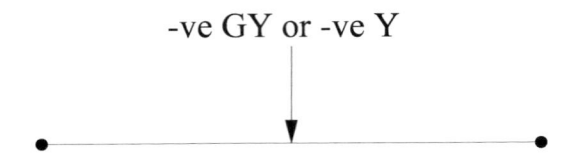

$$\text{-ve GY or -ve Y}$$

- In case of a column as in the illustration below, note that, if you specify the Global X as the direction of the load, the sign of the load will be positive. Alternatively, if you specify the Local Y as the direction of the load, the sign of the load will be negative.

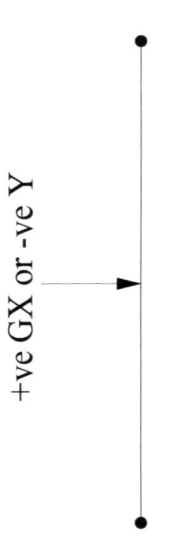

$$\text{+ve GX or -ve Y}$$

- In the case of inclined beam like in the illustration below, note that; the only right choice would be negative Y.

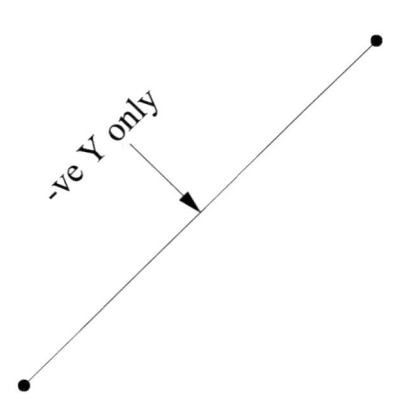

$$\text{-ve Y only}$$

6-12

Module 6: Loading

- For **Porjected** direction, if the Beam was in X-Y plane, then you have only PX, and PY as possible directions for the loads to be applied in the same plane. Accordingly, that applies to the other planes. Also, for Uniform Force, using the **Projected** direction, the value of the load will not be the input value, instead it will be the input value multiplied by the *Sine* or *Cosine* of the angle beween the inclined Beam and the horizontal in the plane of loading.
- To verify, see the following example:
  - If you have the following case (frame in X-Y plane):

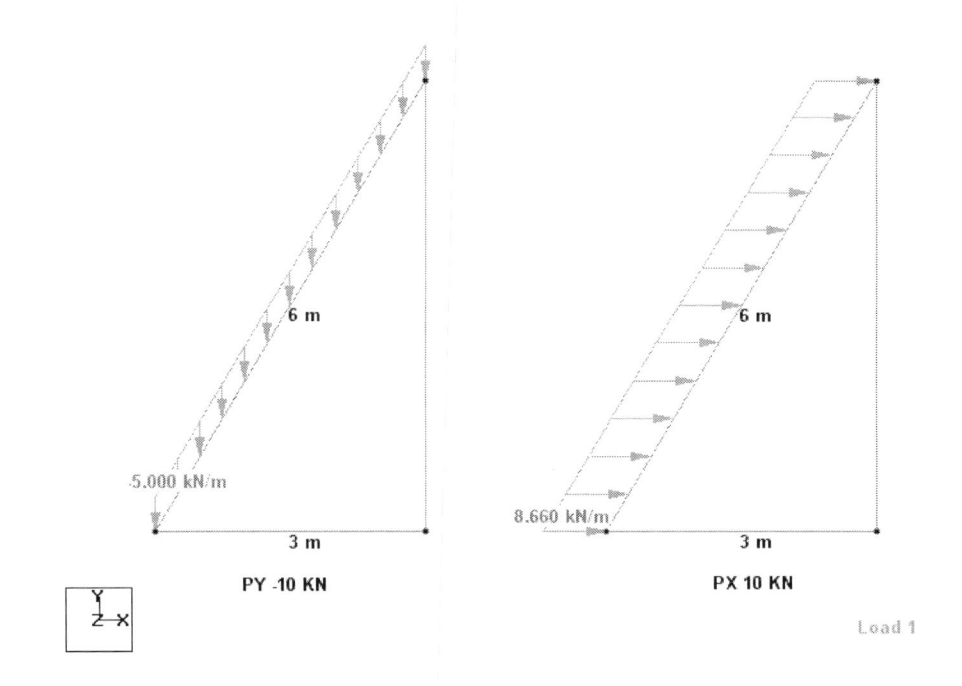

  - The angle between the inclined Beam and the horizontal is 60°.
  - If you apply a force = -10 KN over the 6 m length Beam using PY direction, the value of the load will be -5, why? Because *Cosine*60° = 0.5, and hence -10*0.5=-5.
  - If you apply a force = 10 KN over the 6 m length Beam using PX direction, the value of the load will be 8.66, why? Because *Sine*60° = 0.866, and hence 10*0.866=8.66.

6-13

Creating Primary and Member Loads

Exercise 24

1. Open Common.std.

2. Create Primary Load 1, titled **Dead Load** and include inside it the **Selfweight** only.

3. Create Primary Load 2, titled **Live Load** and create a Uniform Force with value of 25 KN/m (you figure out what is the sign of the load), like the illustration below:

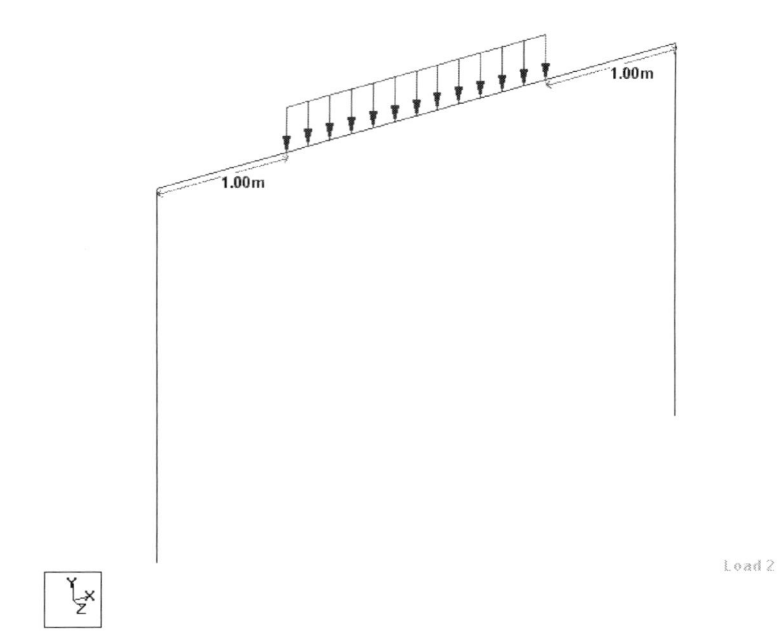

6-14

4. Create Primary Load 3, titled Wind Load and create a Trapezoidal Force with value of W1=10 KN/m, and W2=25 KN/m (you figure out what is the sign of the load), like the illustration below:

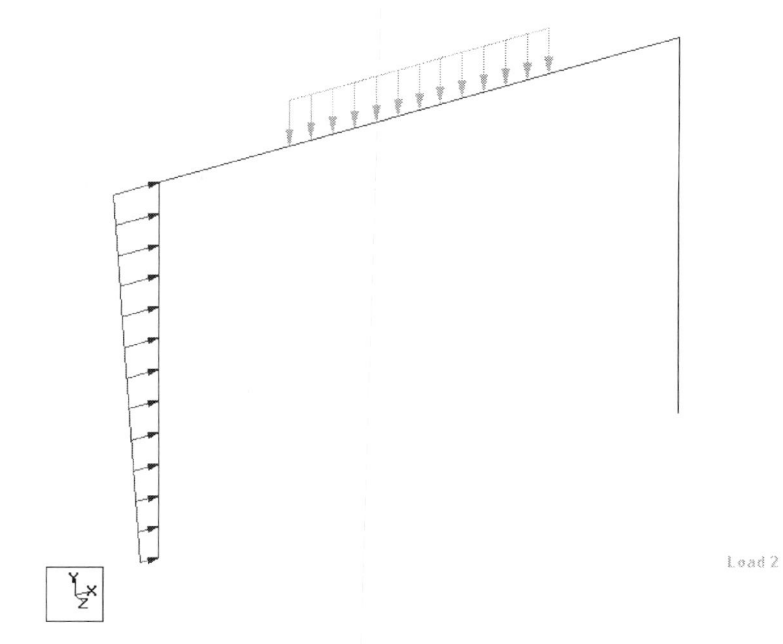

5. Don't save and close.

Linear Varying
- Click Linear Varying tab.
- The case should look like the following:

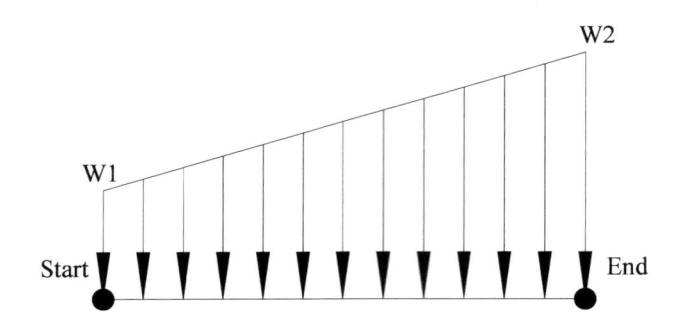

- Or the following case:

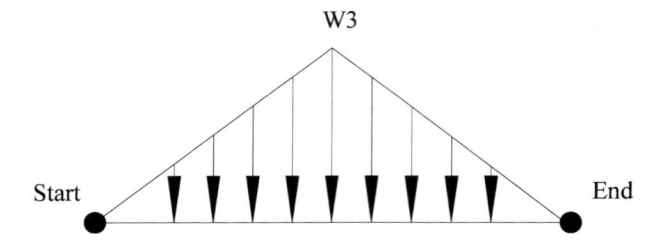

- The following dialog box will appear:

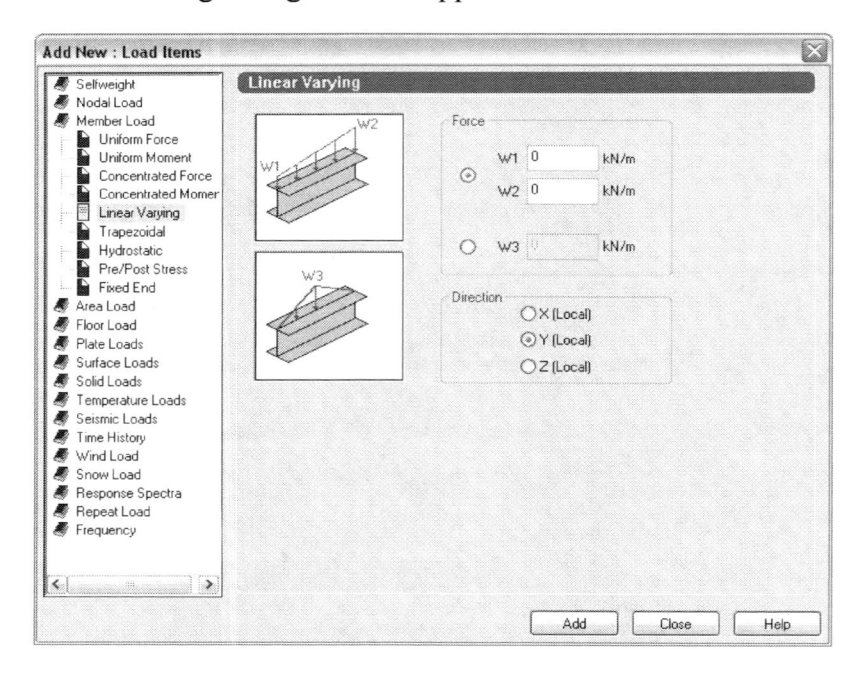

6-16

Module 6: Loading

- Specify **W1**, and **W2** force value, or **W3**:
    - The force should cover the whole Beam length.
    - The loading is always in the Local direction, hence you have only three options rather than nine.

Individual Loads: Area Load

- This load will simulate the One-Way slab loading.
- You don't apply it on a Plate, but on Beams surrounding the Plate.
- STAAD.*Pro* will distribute the pressure value *Force/Area* on the beams according to the theory of One-Way slab.
- Select the desired primary load, and click **Add** button.
- Select **Area Load**, from the left list, the following will appear:

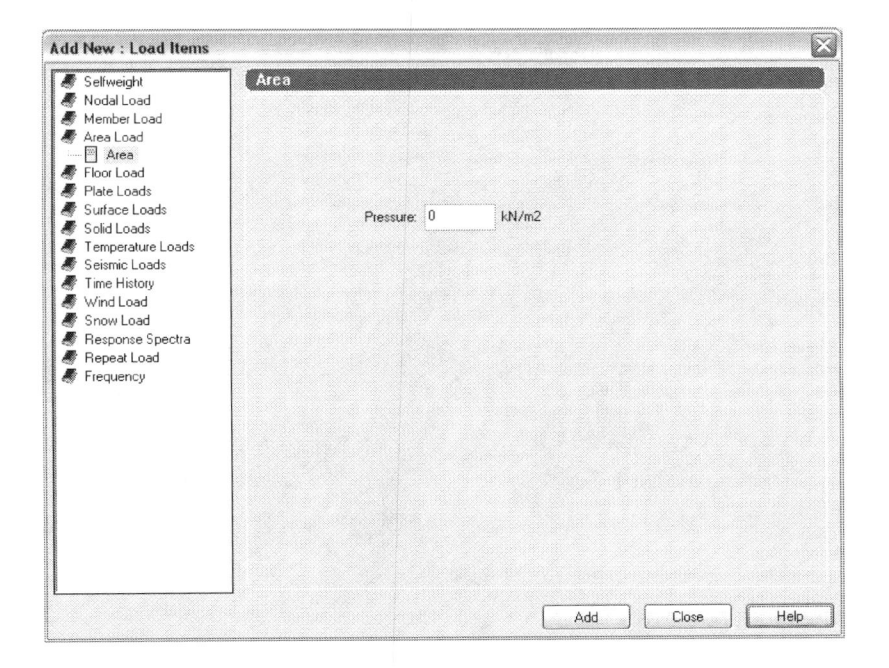

- Specify the pressure value.
- The force will be always in the direction of Local Y.

6-17

Individual Loads: Floor Load

- This load will simulate the Two-Way slab loading.
- You don't apply it on a Plate, but on Beams surronding the Plate.
- STAAD.*Pro* will distribute the pressure value *Force/Area* on the beams according to the theory of Two-Way slab.
- No need to select any Beam before applying this load.

Example
- See the following illustration to understand how the Floor load works:

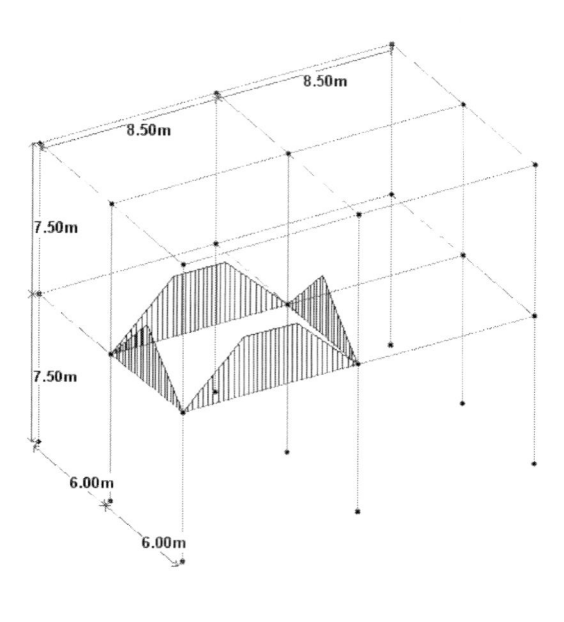

6-18

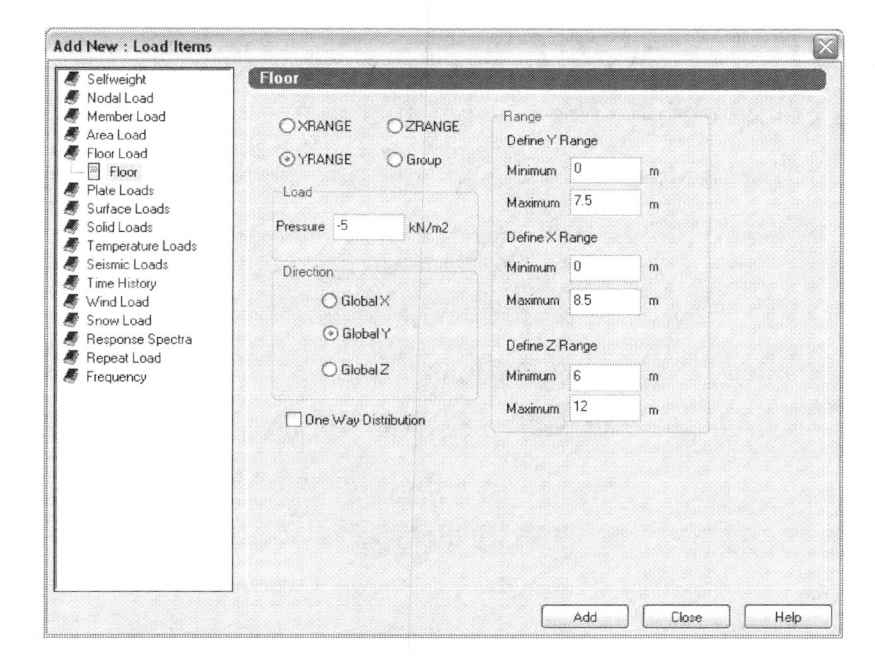

- To create Floor loading on the geometry as illustrated in the above graph, you have to fill the data as in the following dialog box:

Explanation
- Specify the Load value in *Force/Area* units.
- Define the *Y force effectiveness range* which will act in the Global Y direction. In our example the force is affecting the first story of the building and not the second, so range will be from 0 m to 7.5 m.
- Define the *X force effectiveness range* which will act in the Global X direction. In our example the range will be from 0 m to 8.5 m.
- Define the *Z force effectiveness range* which will act in the Global Z direction. In our example the range will be from 6 m to 12 m.

Note
- If you want to load partial part of structure you have to make sure that there are beams at the location you specify, which means, when you say from 6m to 12m, there should be beams at 6m and 12m, or STAAD can't implement the Floor Load.

Individual Loads: Plate Loads

- Plate Loads are loads applied to Plates.
- It is preferable to select the Plates prior to invoking the command.

6-steps procedure
- This is a very simple approach to applying load to Plates:
  - Make sure that you are showing the Plate Orientation (right-click, then select **Labels**, under **Plates**, click **Plate Orientation** on).

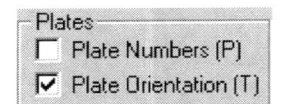

  - Select the Plate(s).
  - Select the desired primary load, then click **Add** button.
  - Click the **Plate** button. A dialog box will appear.
  - Specify the type, value of the load.
  - Specify the direction of the load, then click **Add** button.
  - From Assignment method select **Assign to Selected Plates**, then click **Assign** button.

Pressure on Full Plate
- Click Pressure on Full Plate.
- The pressure will be *Force/Area* covering the whole Plate.
- The following dialog box will appear:

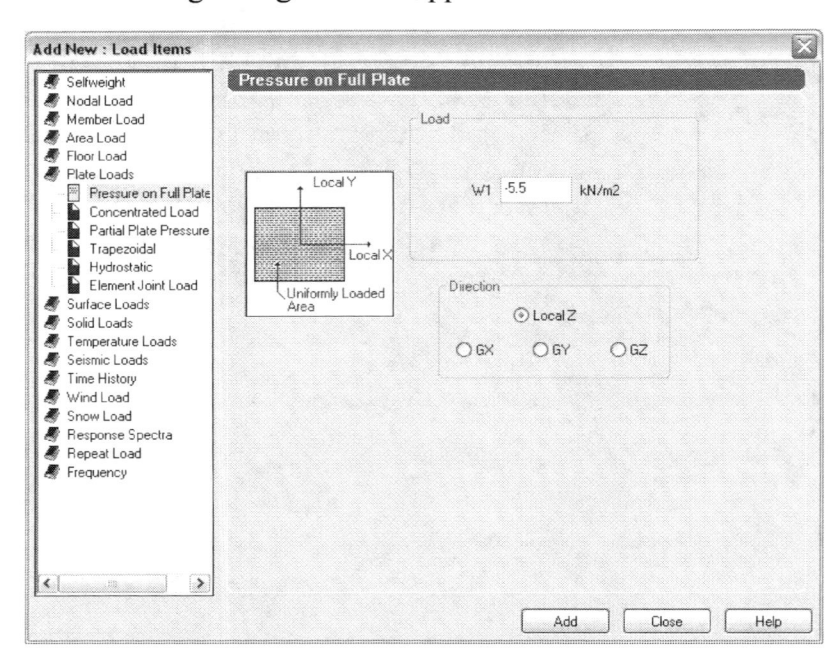

- Specify the pressure value W1.
- Specify the Direction of the pressure. You have four choices:
  - Local Z (Prependicular on the Plate), don't forget the negative sign so the pressure will be acting downwards.
  - GX, GY, and GZ, the three global direction, and accroding to the direction specify the sign of the load.
- The result may look like the following:

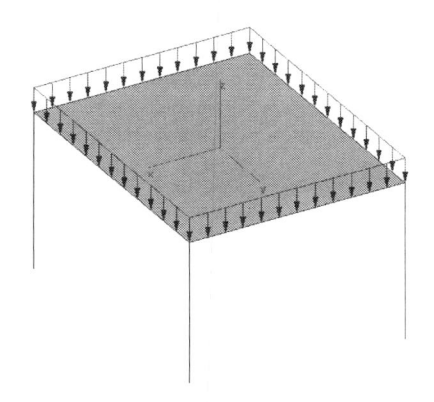

Note
- Although it looks like the load is on the Beams, but actually the load is acting on all the Plate.

Partial Plate Pressure Load
- Click Partial Plate Pressure Load.
- The pressure will be *Force/Area* covering the Plate partially. The following dialog box will appear:

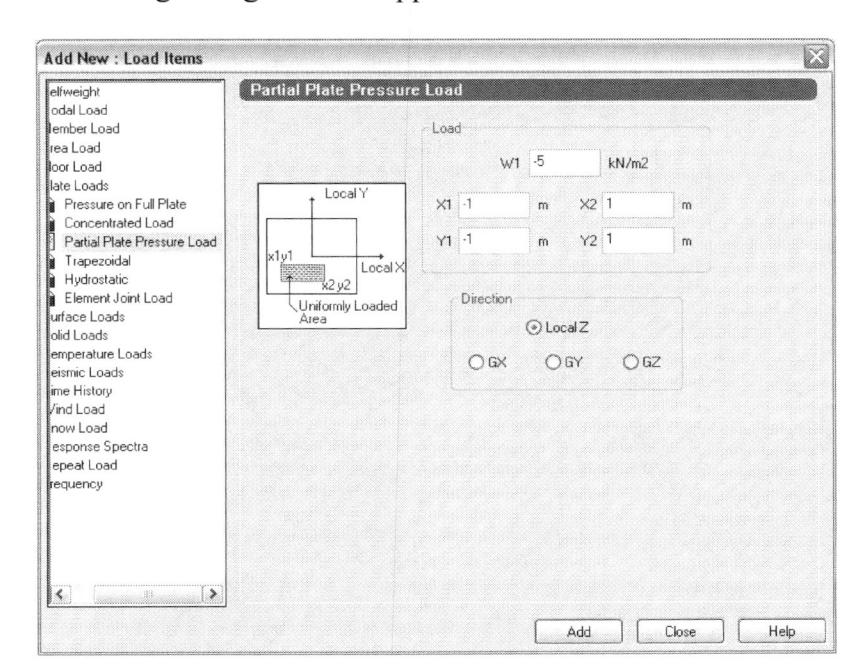

6-21

- Specify the pressure value **W1**.
- Specify X1, Y1 which are the coordinates of the start point of the area to be covered, measured from the centroid of the Plate.
- Specify X2, Y2 which are the coordinates of the end point of the area to be covered, measured from the centroid of the Plate.
- If you put X1 = -1, Y1 = -1, this means 1 m to the left and below the centroid. Also, if X2 = 1, Y2=1 that means 1 m to the right and above the centroid.
- Based on X1, Y1, X2, Y2 mentioned above the area to be covered will be 2X2 m^2.
- Specify the direction just like in the **Pressure on Full Plate**.

- The result may look like the following:

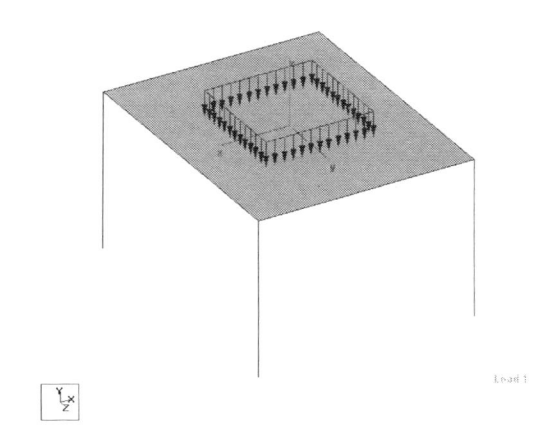

Module 6: Loading

Concentrated Load
- Click Concentrated Load.
- The following dialog box will appear:

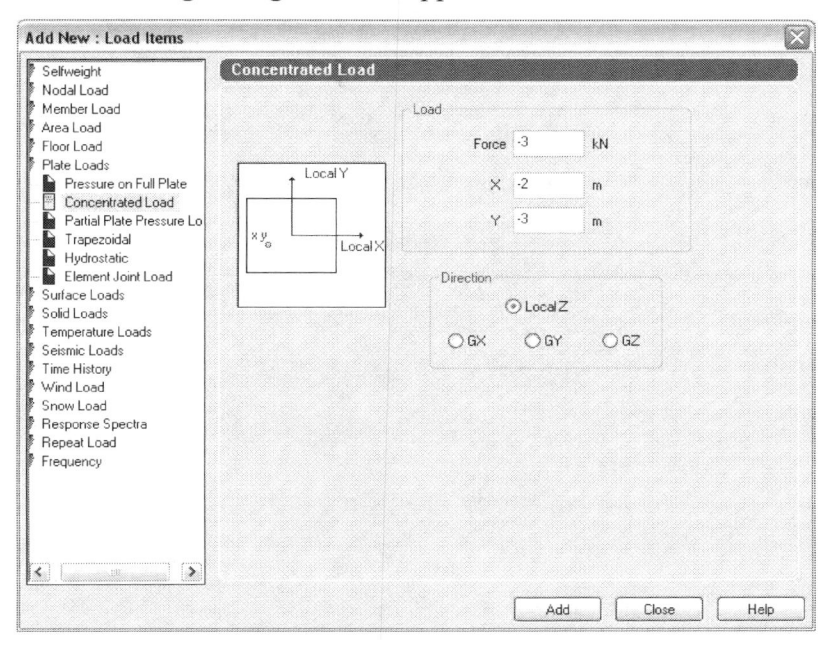

- Specify the **Force** value.
- Specify the location of the concentrated load by specifying the X, and Y coordinated measured from the centroid of the plate.
- Specify the direction just like in the **Pressure on Full Plate**.

- The result may look like the following:

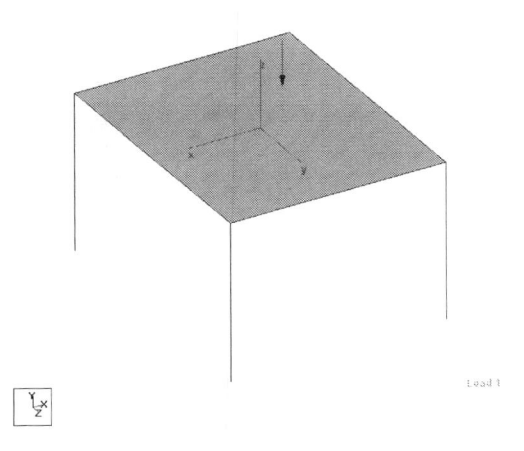

Trapezoidal
- Click Trapezoidal.
- The pressure will cover the whole Plate.
- The following dialog box will appear:

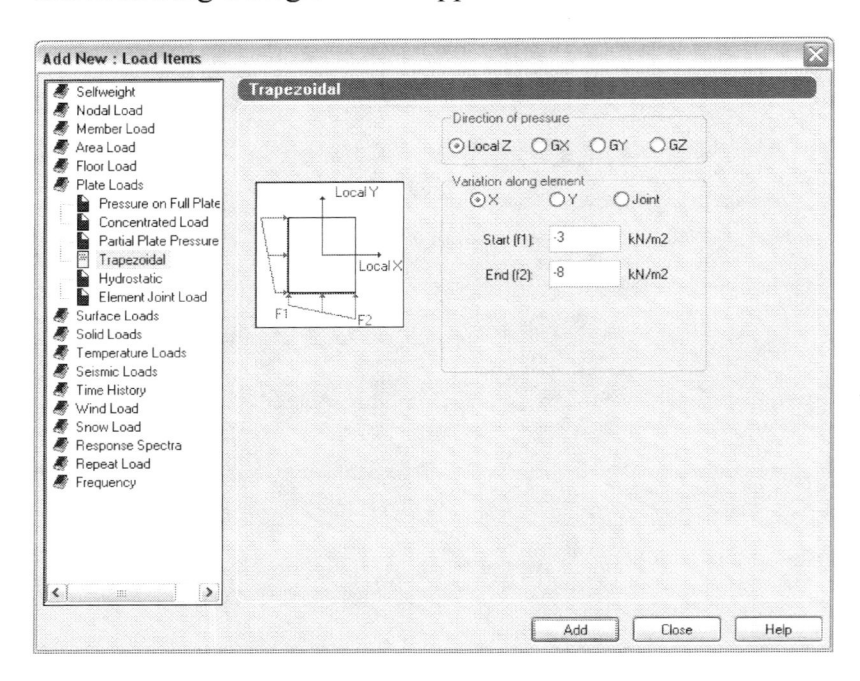

- Specify the Direction of the Pressure.
- Specify the value of Start force f1, and End force f2.
- Specify the **Variation along Element**, in which the pressure will vary along it. The illustration below will show a pressure will change along the local X.

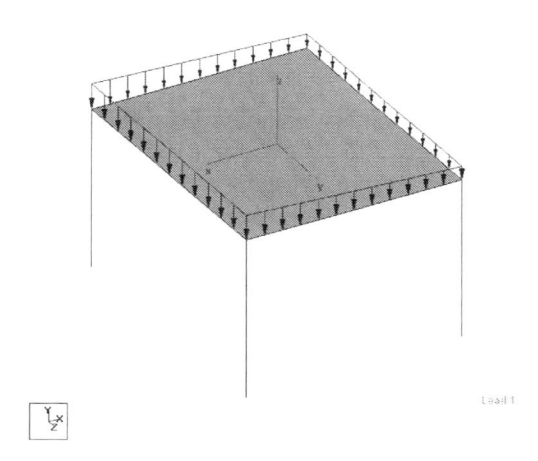

6-24

Individual Loads: Node Loads

- Node Loads are loads applied to Nodes.
- Select the Nodes.
- Select the desired primary load case, and click **Add**, a dialog box will appear. Click the **Nodal Load** then select **Node**.
- Specify the value and the sign of the load, then click **Add** button.
- From Assignment method select **Assign to Selected Nodes**, then click **Assign** button
- The following dialog box will appear:

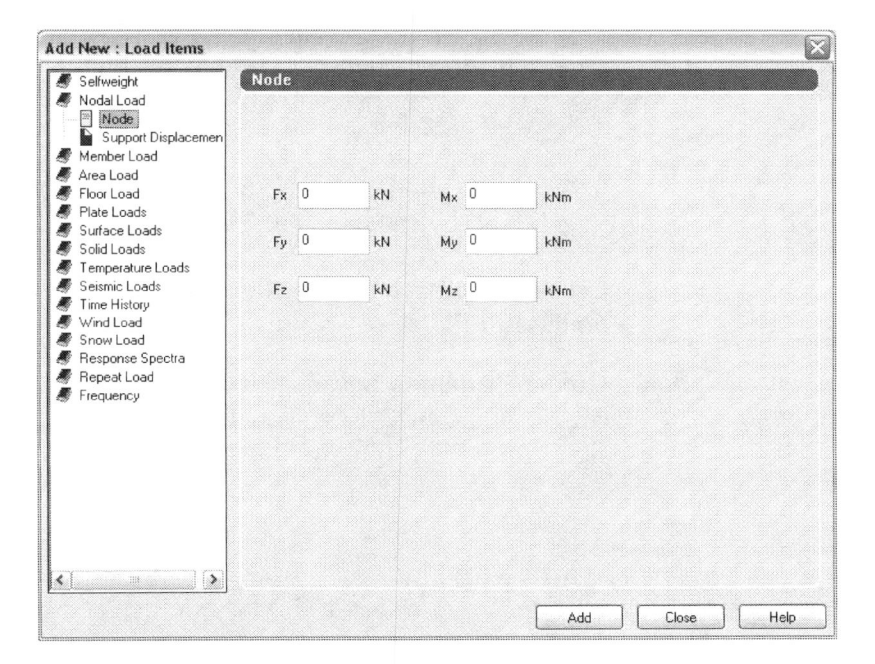

- The illustration below will show the force working in the negative Z axis.

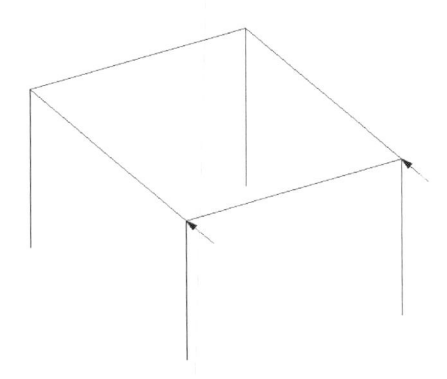

Creating Floor, Plate, and Nodal Loads

Exercise 25

1. Start STAAD.*Pro*, and create 3D-frame **4m** in X-axis, **3m** in Y-axis, and **5m** in Z-Axis. Add a single plate on the top of the frame.

2. Create Primary Load 1, titled Dead Load, which includes:

    a. Selfweight.

    b. Floor Load = -5 KN/m^2, which will cover the one story structure. (Hint Y Range = 0 to 3)

3. Create Primary Load 2, titled Live Load, which includes Pressure Load on the full Plate = -3 KN/m^2.

4. Create Primary Load 3, titled Wind Load, which includes Nodal Load with a value of 2 KN (you figure out which direction and what is the sign of the load) like the illustration below:

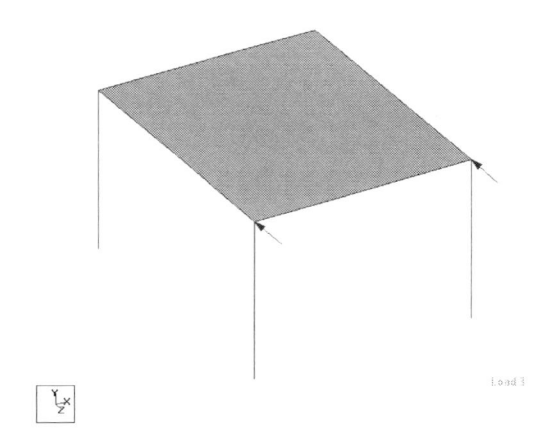

5. Don't close the file you will need it in the next exercise.

Individual Loads: Viewing & Editing

Viewing
- To view the load value and Floor Load, right-click and select **Labels**. The Labels dialog box will appear, under the part of **Loading Display Options**:
  - To view the Load Value on the screen click **Load Values** ON.
  - To view the Floor Loading on the Beams, click **Display Floor Loading** ON.
  - To view the two trinagles and the two trapezoidal of the Two-Way slab loading click **Display Floor Load Distribution** ON.

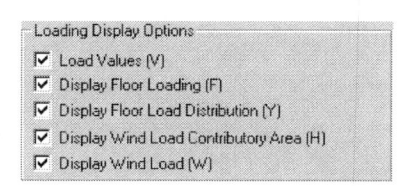

- The following illustration clarifies more by displaying the Load Value, Floor Loading, and finally Floor Loading Distribution.

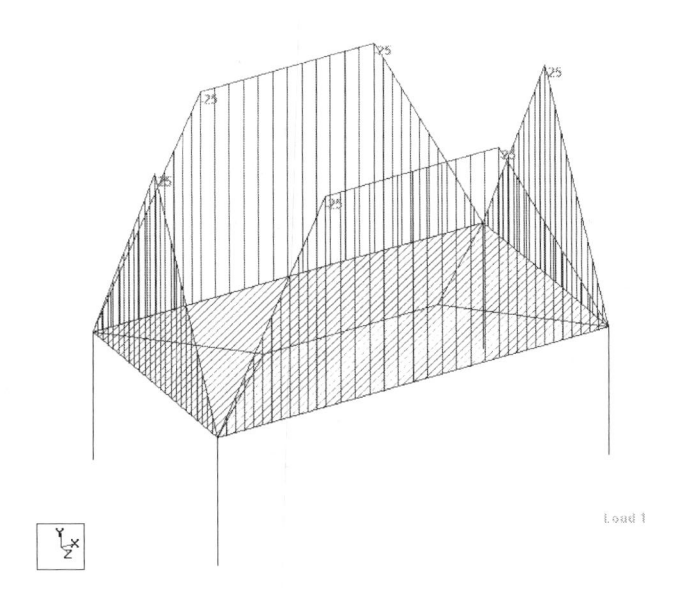

Editing ■ When you are in the sub-page **Load** the Data Area would look like the following:

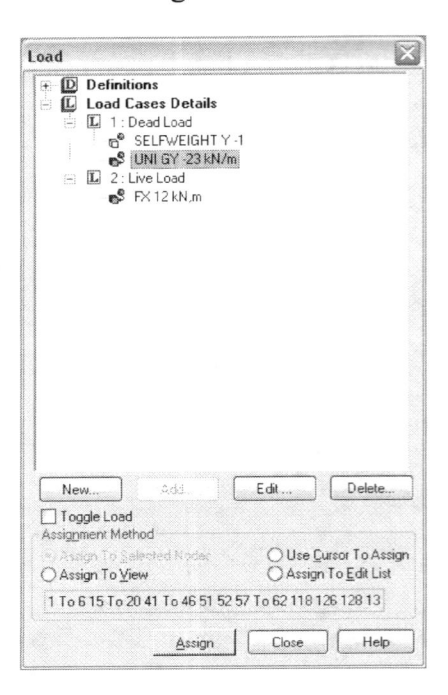

■ Each load (Selfweight, Member, Plate, or Nodal) will be displayed in this part of the screen, after selecting the load from the list you can:

- Delete the individual load, by clicking **Delete** button, or clicking the **Del** key on the keyboard.

- If you click **Edit** button, or if you double-click the load from the list a dialog box will appear to give you the chance to edit the value of the load, and its direction (if applicable).

- From the **Selection** toolbar, select **Load Edit Cursor** icon, select the load itself (and not the Node, Beam, or Plate), to show the dialog box of the load to edit it. Only Area Load, and Floor Load can't be edited in the last way.

How to Create Manual Combinations

- Manual Combination is any combination:
  - User will create it, and not STAAD.*Pro*
  - User will specify the factors to be multiplyed by the Primary Loads.
  - User will specify how many combinations are needed.

- To create a combination, click on **Load Cases Details** at the Data Area, then click **Add** button, a dialog box will appear, click **Define Combinations**, you will see the following:

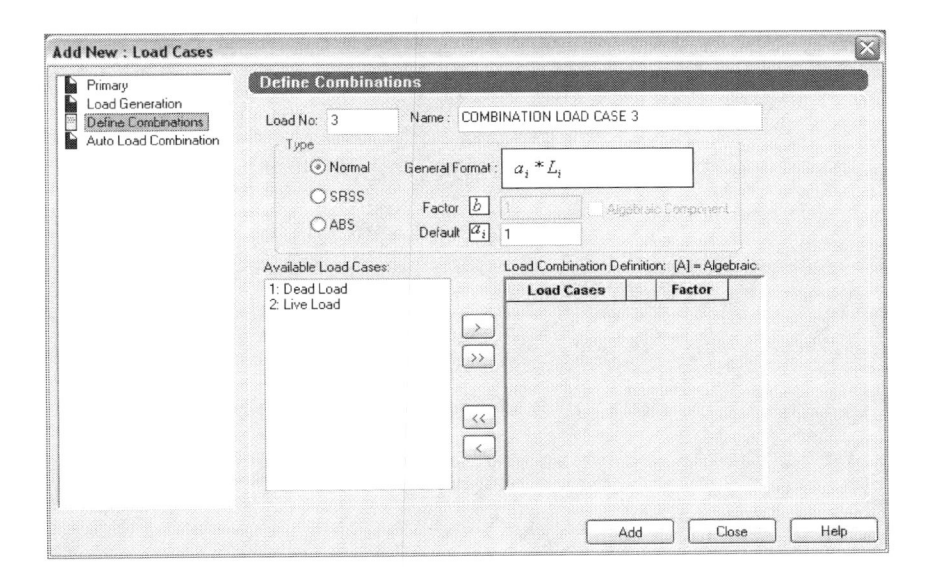

[Dialog box: Add New: Load Cases — Define Combinations. Load No: 3, Name: COMBINATION LOAD CASE 3. Type: Normal / SRSS / ABS. General Format: $a_i * L_i$. Factor b, Default a_i = 1. Available Load Cases: 1: Dead Load, 2: Live Load. Load Combination Definition: [A] = Algebraic.]

- As you can see STAAD.*Pro* gave the first Combination a number which will be a continuation of the last Primary Load number. Type the Name of the combination (remember the name doesn't mean any thing to STAAD.*Pro*). Make sure that the Type is **Normal**.
- In the **Default** field type the desired factor (1.4, 1.6, or any other factor).
- Now select the primary load case, and click <u>one arrow to the right</u>.
- Repeat the same process for the other primary loads.

From Menus
- From the menus, select Commands/Loading/Load Combination.
- This command will not take you to the **Load** sub-page, but instead will show you the same dialog box shown above, start with **New** button to create a new Combination.

How to Create Automatic Combinations

- Automatic Combination is any combination:
    - Created by STAAD.*Pro* according to specific codes.
    - STAAD.*Pro* will specify the factors which will be multiplyed by the Primary Loads.
    - STAAD.*Pro* specify how many combination is needed.
- As pre-requisite user should take care of the following points:
    - When creating a new Primary Load, select **Loading Type** from the pop-list, which contians: Dead, Live, Roof Live, Wind, Seismic, Snow, Fluids, etc. Don't mix up the **Loading Type** with the **Title**. Check the following dialog box:

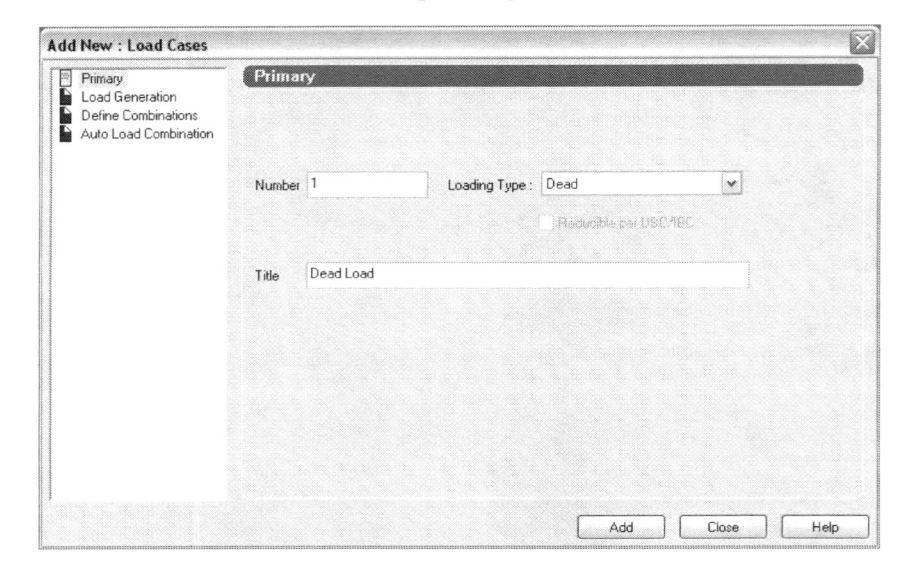

- As you can see we selected Primary Load 1 to be **Dead**. Do the same for 2 to be Live, and 3 to be Wind (just for the sake of an example).

Module 6: Loading

- When you are done with Primary Loads, create a combination by clicking on **Load Cases Details** at the Data Area, then click **Add** button, a dialog box will appear, click **Auto Load Combinations**, you will see the following dialog box:

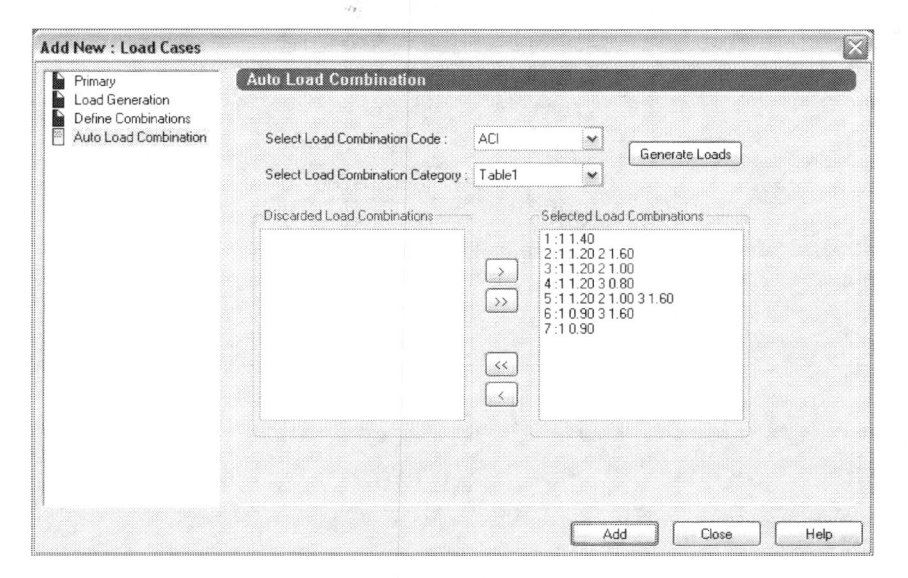

- Select the desired code, then select the **Load Combination Category** which will decitate the number of combination the program will generate.

- Now click **Generate Loads** button. Number of combination will be generated automatically.

- Once you are done, click **Add** button.

6-31

Editing Primary Loads and Creating Combinations

Exercise 26

1. Continue with the previous file.
2. Change the value of the Floor Load from -5 to -2.5
3. View the Load Values.
4. Create the following manual Combinations:
    a. 1.4 DL + 1.6 LL
    b. 0.75(1.4 DL + 1.6 LL + 1.6 WL), which is 1.05DL+1.2LL+1.2WL

Creating Primary Loads & Combinations (Concrete Structure)

Workshop 4-A

1. Open Small_Building file.
2. Create a Primary Load 1 Type = Dead, titled **Dead Load**, contains:
   a. Selfweight (Y, -1).
   b. Floor Loading -5 KN/m^2, YRange=0-6, XRange=0-10, ZRange=0-5.
3. Create a Primary Load 2 Type = Live, titled **Live Load**, contains: Floor Load -8 KN/m^2, YRange=0-6, XRange=0-10, ZRange=0-5
4. Create a Primary Load 3, titled **Wind Load**, contains: Nodal Load over the Nodes as shown Fx = 5 KN:

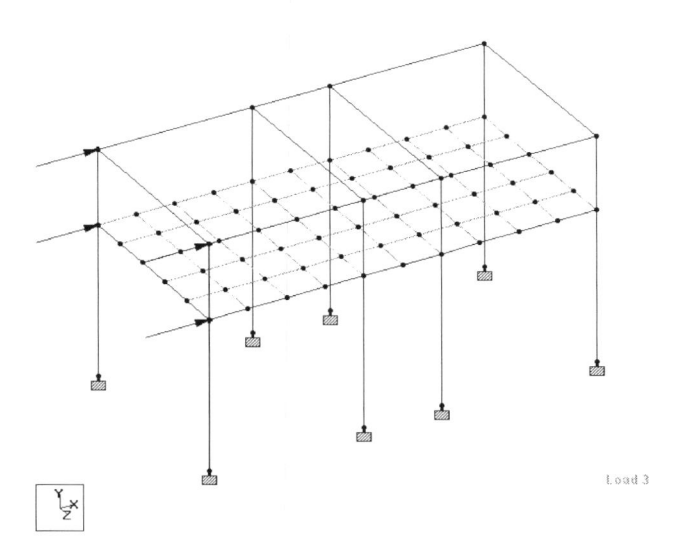

5. Create automatic combinations using **ACI** code, and **Table 1** category.
6. Save and Close.

Creating Primary Loads & Combinations (Steel Structure)

Workshop 4-B

1. Open Small_Building file.
2. Create a Primary Load Number 1 Type = Dead, titled **Dead Load**, contains:

    a. Selfweight (Y, -1).

    b. Uniform Force=-10 on the Beams as shown.

    c. Nodal Load = -55 KN as shown:

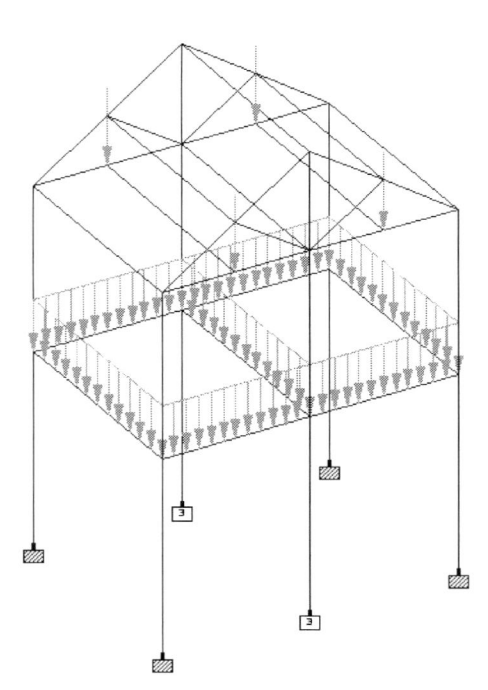

3. Create a Primary Load Number 2 Type = Live, titled **Live Load**, contains:

   a. Uniform Force = -15 KN/m on the top chord as shown (Use Y).

   b. Uniform Force = -20 KN/m on the beams as shown.

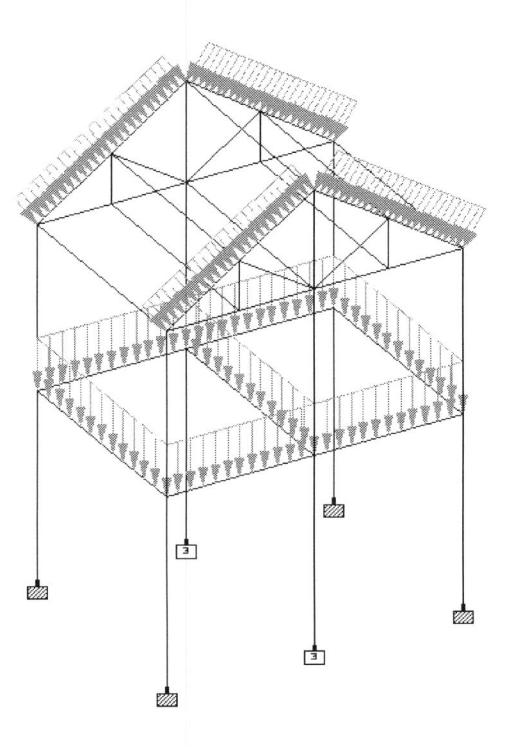

4. Create a Primary Load Number 3 Type = Wind, titled **Wind Load**, contain:

   a. Uniform Force = 13 KN/m on th left columns, as shown.

   b. Uniform Force = -17 KN/m of the left top chord as shown (use Y).

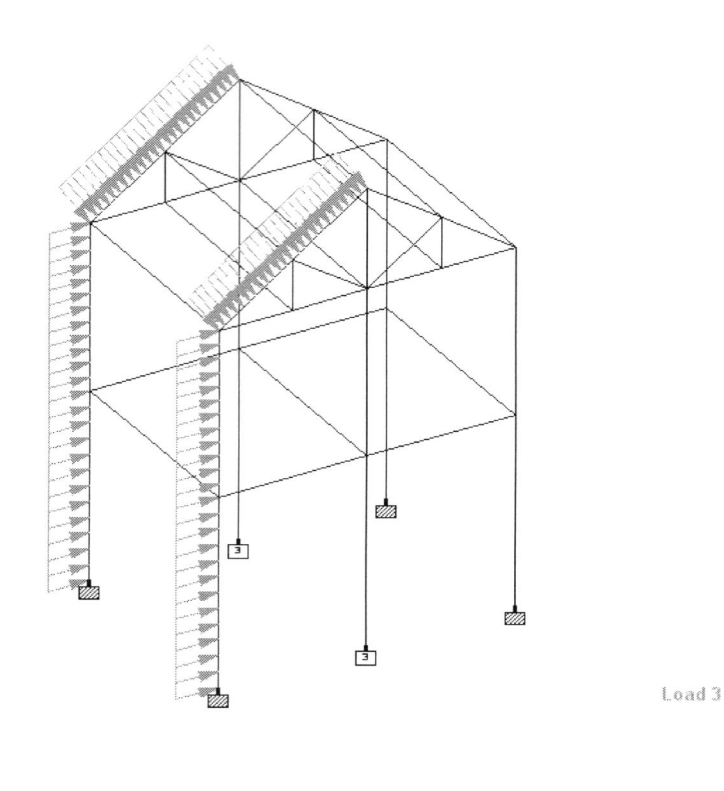

5. Make automatic combinations, using **AISC** code, and **Garage** Category.
6. Save and Close.

Module Review

1. Combinations is to simulate Design Codes Combinations:
    a. True
    b. False

2. Which of the following is true about GX, GY, and GZ:
    a. Used to specify the direction of load.
    b. The only 3 methods available to specify the direction of the load.
    c. You can't use Global direction to specify the direction of the load.
    d. None of the above.

3. Primary Load includes _____ Loads.

4. You can't edit the Floor Load from the Data Area:
    a. True
    b. False

5. Which is true about Plate Load:
    a. Floor load not among them
    b. Load Applied on Plates
    c. Loads applied to the Beams surrounding the Plate
    d. Answers a & b

6. After you define a Primary Load, the Load number will appear in ___ different places.

Module Review Answers

1. a
2. a
3. Individual
4. b
5. d
6. 3

Module 7:

Analysis

This module contains:

- Types of Analysis
- Perform Analysis Command
- P-Delta Analysis Command
- Non-Linear Analysis Command

Introduction

- STAAD.*Pro* Analysis can be:
  - Static Analysis
  - Dynamic Analysis
- Static Analysis can be either:
  - Perform Analysis (Linear Analysis)
  - P-Delta Analysis
  - Non-Linear Analysis
- Dynamic Analysis can be either:
  - Time History
  - Response Spectrum
- In this tutorial we will cover only the types of the Static Analysis commands, and their requirements.

Note
- Analysis command is a line, which will be added to your input file, and ***not*** the execution command. Analysis command will tell STAAD.*Pro* what is the desired analysis *type* you want to use to get your results.

Perform Analysis Command

- The first analysis command is **Perform Analysis**.
- When you use this command STAAD.*Pro* would understand the following:
  - Displacement of Nodes is so minimal that will not cause any secondary loading.
  - Deflection of Beams is so minimal that will not cause any secondary loading.
  - The structure will be analyzed once.
  - The results of the first iteration will be considered the Analysis results.

- Select **Commands/Analysis/Perform Analysis**, the following dialog box will appear:

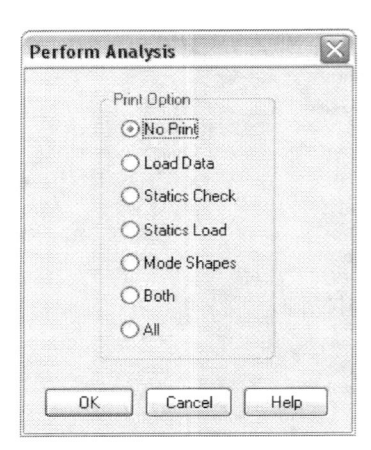

- Select one of the **Print Options**, and click **OK**.
- Or from the Page Control select **Analysis/Print** the following dialog box will pop-up automatically:

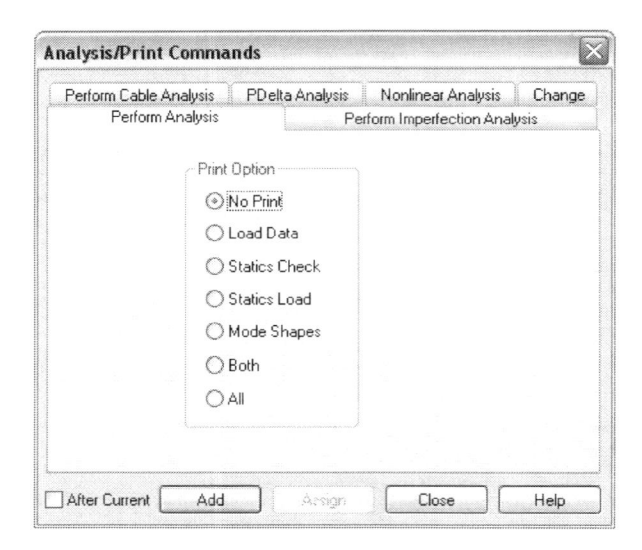

Module 7: Analysis

- Select one of the Print Options, and click **Add**, then **Close**.
- In both cases the line of the Analysis type will be added to the input file, and you can see that in the Data Area as follows:

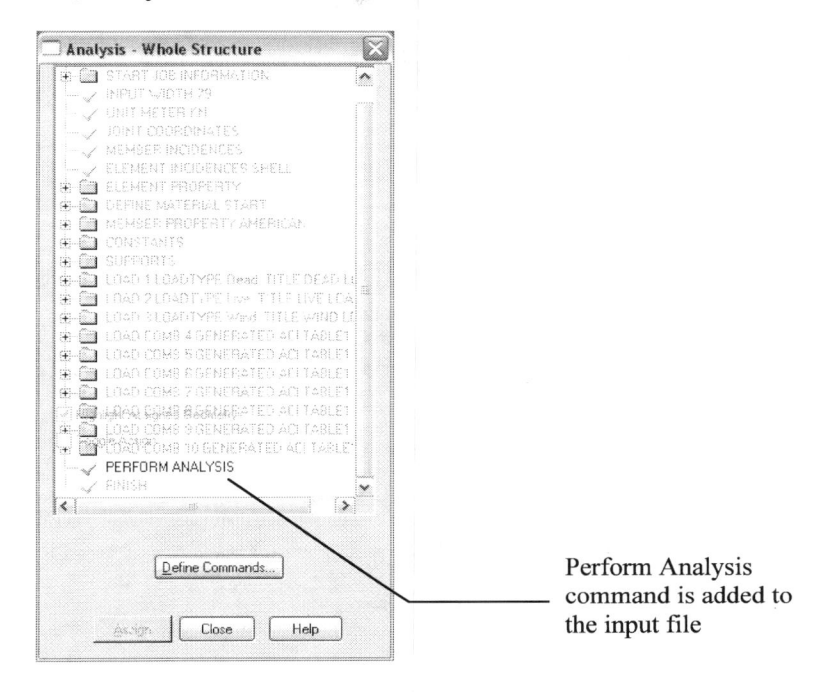

Perform Analysis command is added to the input file

- The **Print Options** in both dialog boxes are not related to the analysis results generation, but instead they are some add-on print option for loads and static check:

No Print
- Do the Analysis and don't print any additional information.

Load Data
- Plus to Analysis, it'll print the Primary Load cases, which of the Nodes/Beams/Plates the Primary Loads are affecting, the value, and the location.

Statics Check
- Plus to Analysis, it will print two sets of summations:
  - The Σ of applied loads, and moments around center of gravity.
  - The Σ of reaction loads, and moments around center of gravity.
- Also, it'll print the max displacement as movement and rotation.

Statics Load
- Like Statics Check, plus external and internal Joint Loads for supports.

Mode Shape
- Only for Dynamic Loading

Both
- Load Data + Statics Check.

All
- Load Data + Statics Check + Statics Load

P-Delta Analysis Command

- The first of the two second-order stability analysis.
- A multi-iterative analysis.
- Check the following case of loading:

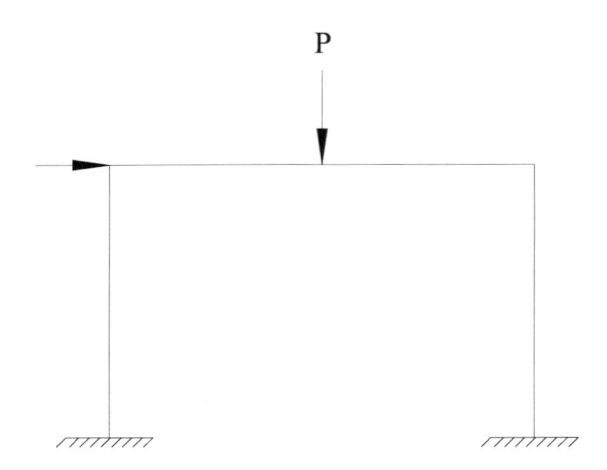

- If the lateral load and the vertical load are working simultaneously, the P-Delta Analysis will go through the following steps:
  - Calculate the Primary displacement of Nodes based on the external loads (as discussed in the Perform Analysis).
  - Due to the Displacement of the Nodes, STAAD.*Pro* will calculate the second order loadings.
  - P will be revised to show the new value of loading (the original external load + secondary loading).
  - The location of P will be revised also.
  - The new system is ready for another iteration.

Module 7: Analysis

- The **Number of Iterations** specified by the user, and/or the **Convergence** value will stop the analysis engine from doing another iteration.
- **Convergence** is the tolerance value, which the user will set as a reference, if the displacement of Nodes is less than this value the P-Delta Analysis will stop.
- The new system would look something like:

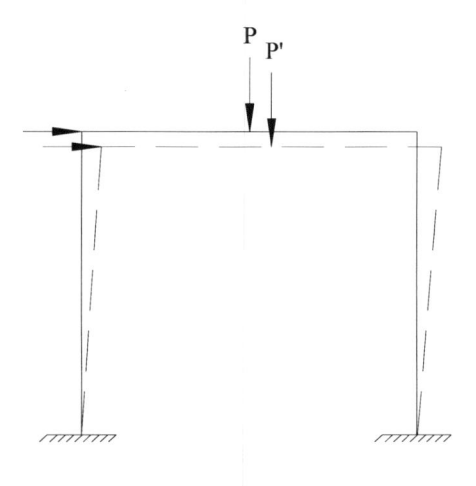

- Select **Commands/Analysis/P-Delta** Analysis, the following dialog box will appear:

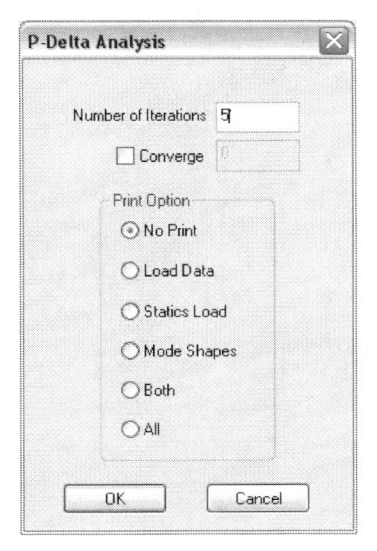

STAAD.Pro 2005 Tutorial

- Specify the Number of Iterations.
- Click on the **Converge** value and Specify it.
- Select **Print Options**, and click **OK**.
- Alternatively, from the Page Control select **Analysis/Print** the following dialog box will pop-up automatically:

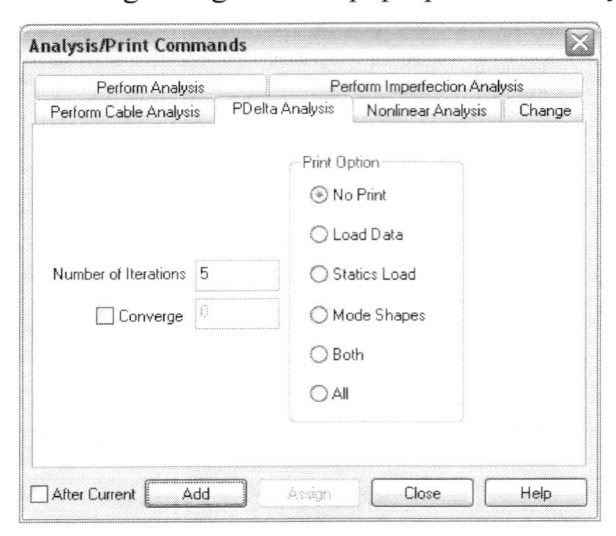

- Click on the **PDelta Analysis** tab, and do as in the previous method, click **Add**, then **Close**.
- In both cases the line of the Analysis type will be added to the input file, and you can see that in the Data Area as follows:

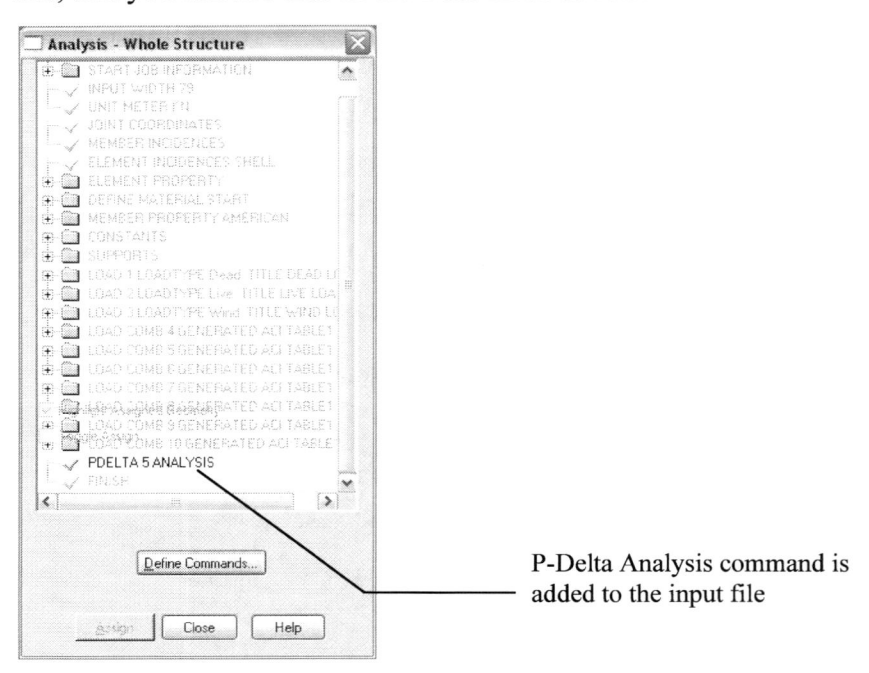

P-Delta Analysis command is added to the input file

7-8

Note
- Several codes like ACI 318, LRFD, and IS456-1978 recommends the using of the P-Delta Analysis instead of the Perform Analysis.
- Because Perform Analysis command will not generate exact values for the shear/moment, but instead approximate values, hence user should magnify the moments by a factor to compensate the approximation.
- Only applied on Beams, and not Plates.

Non-Linear Analysis Command

- With P-Delta Analysis STAAD.*Pro* took care of the secondary loading caused by the Displacement of the Nodes. But, deflection of Beams was not considered.
- Non-Linear Analysis takes care of both secondary loading caused by Displacement of the Nodes, and geometric stiffness correction caused by Deflection of Beams.
- Suitable for long spans beams with high deflection.
- Deflection in considered in the Analysis, rather than to be checked for in the design.
- A multi-iterative analysis.
- Check the following case of loading:

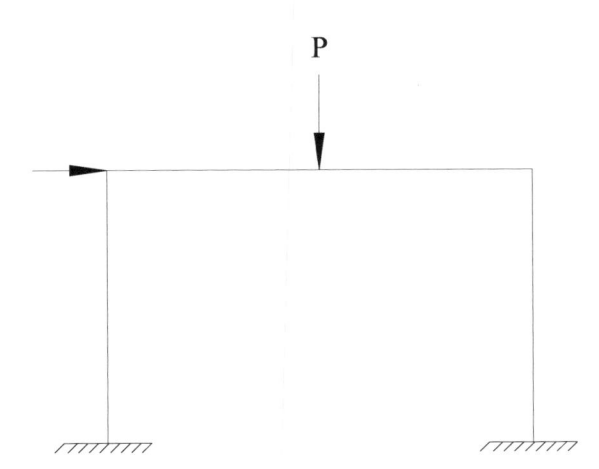

- The Non-Linear Analysis will go through the following steps:
  - Calculate the Primary displacement of Nodes based on the external loads (as discussed in the Perform Analysis).
  - Due to the Displacement of the Nodes, STAAD.*Pro* will calculate the second order loadings, and due to Beam Deflection STAAD.*Pro* will calculate the stiffness correction.
  - P will be revised to show the new value of loading (the original external load + secondary loading + stiffness correction).
  - The location of P will be revised also.
  - The new system is ready for another iteration.
- The **Number of Iterations** specified by the user, will stop the analysis engine from doing another iteration.
- The new system would look something like:

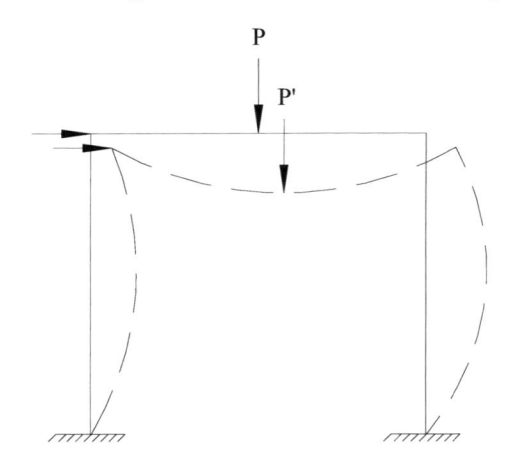

Module 7: Analysis

- Select **Commands/Analysis/Non-Linear Analysis**, the following dialog box will appear:

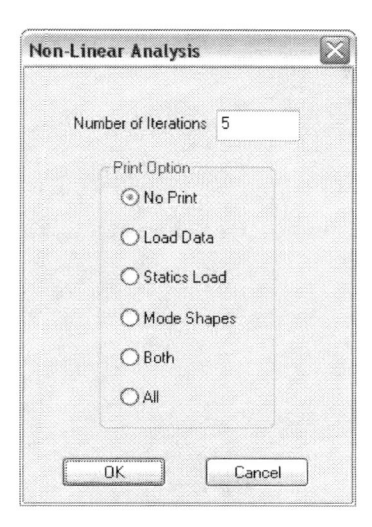

- Specify the Number of Iterations.
- Select **Print Options**, and click **OK**.
- Alternatively, from the Page Control select **Analysis/Print** the following dialog box will pop-up automatically:

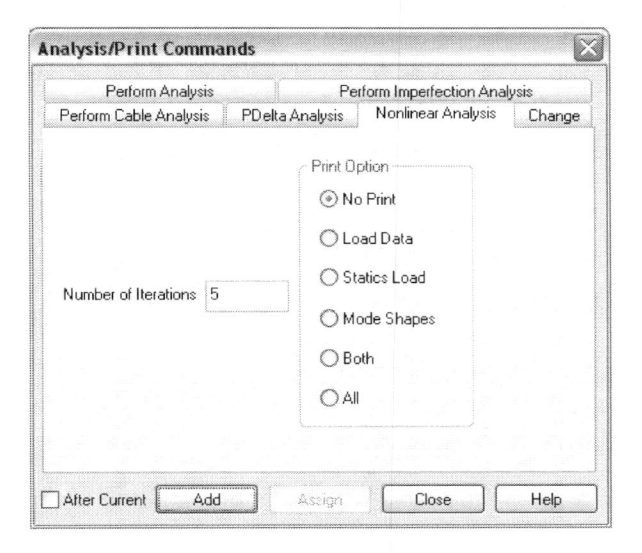

- Click on the **Nonlinear Analysis** tab, and do as in the previous method, click **Add**, then **Close**.

- In both cases the line of the Analysis type will be added to the input file, and you can see that in the Data Area as follows:

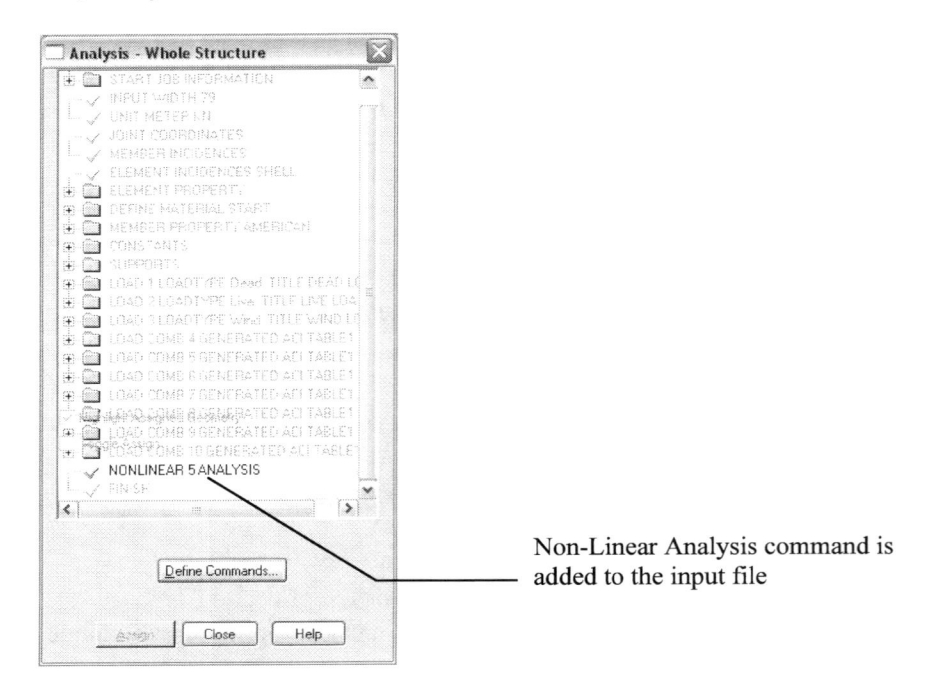

Non-Linear Analysis command is added to the input file

Note ■ P-Delta Analysis, and Non-Linear Analysis will work only with stuctures that can transfer the axial loads from one story to another.

The Execution Command

- Once the Analysis command is added to the input file, it will be ready for execution.
- From the menus select **Analyze/Run Analysis**, the following dialog box will appear:

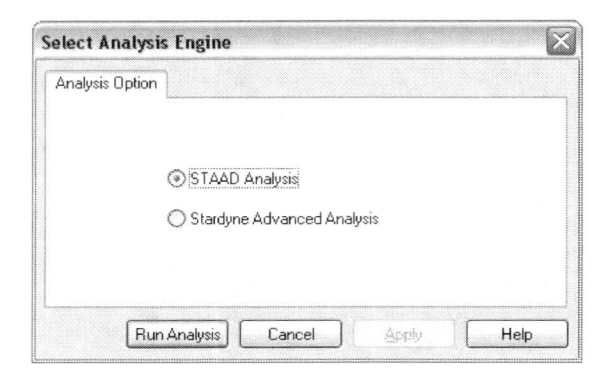

Module 7: Analysis

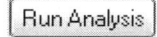

[Run Analysis]
- Make sure that **STAAD Analysis** is selected, and click **Run Analysis** button.
- Now the entire input file will be sent to the STAAD.*Pro* Analysis engine.
- STAAD.*Pro* Analysis engine will read the input file from left to right, and from top to bottom, verifying the following:
  - All the information to formulate a complete structure is there (nothing is missing, e.g. one of the supports was not defined)
  - The structure, which the input file describes, is stable structure.
  - STAAD.*Pro* syntax has been followed, for instance, before the Node coordinates, the input file should include **Joint Coordinate** statement. If this statement was misspelled, or absent, STAAD.*Pro* will give a syntax error.
- That doesn't guarantee the correctness of the results.

First Case
- If everything seems to be OK, STAAD.*Pro* will produce something like the following dialog box:

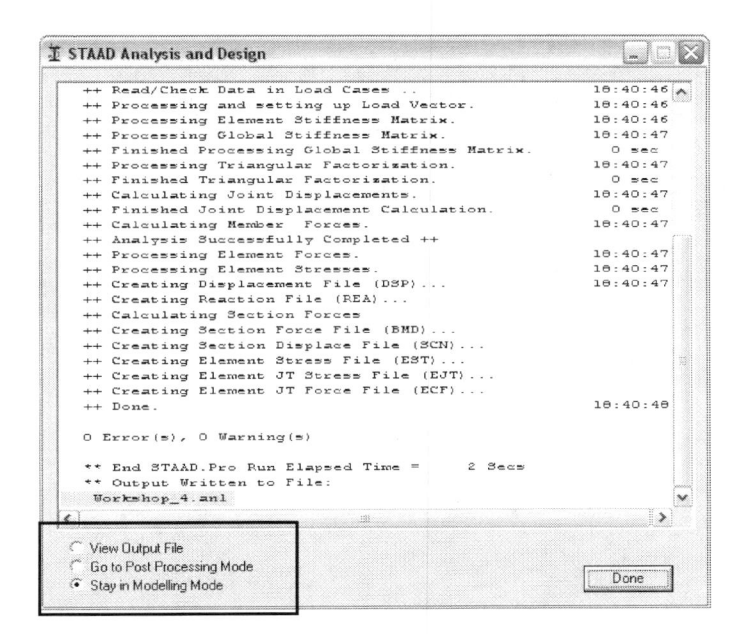

```
STAAD Analysis and Design
++ Read/Check Data in Load Cases ..              18:40:46
++ Processing and setting up Load Vector.        18:40:46
++ Processing Element Stiffness Matrix.          18:40:46
++ Processing Global Stiffness Matrix.           18:40:47
++ Finished Processing Global Stiffness Matrix.     0 sec
++ Processing Triangular Factorization.          18:40:47
++ Finished Triangular Factorization.               0 sec
++ Calculating Joint Displacements.              18:40:47
++ Finished Joint Displacement Calculation.         0 sec
++ Calculating Member  Forces.                   18:40:47
++ Analysis Successfully Completed ++
++ Processing Element Forces.                    18:40:47
++ Processing Element Stresses.                  18:40:47
++ Creating Displacement File (DSP)...           18:40:47
++ Creating Reaction File (REA)...
++ Calculating Section Forces
++ Creating Section Force File (BMD)...
++ Creating Section Displace File (SCN)...
++ Creating Element Stress File (EST)...
++ Creating Element JT Stress File (EJT)...
++ Creating Element JT Force File (ECF)...
++ Done.                                         18:40:48

0 Error(s), 0 Warning(s)

** End STAAD.Pro Run Elapsed Time =     2 Secs
** Output Written to File:
   Workshop_4.anl

  ○ View Output File
  ○ Go to Post Processing Mode
  ● Stay in Modelling Mode                    [ Done ]
```

- In this dialog box STAAD.*Pro* will list each step is going through and the time taken to accomplish it.
- Then it will show the following statement: "0 Error(s), 0 Warning(s)".

7-13

- STAAD.*Pro* will report that it created the following files:
  - Displacement file, *filename.dsp*
  - Reaction file, *filename.rea*
  - Section Force file, *filename.bnd*
  - Section Displacement file, *filename.scn*
- STAAD.*Pro* will give the user three options to choose the next step:
  - To **View Output File**. With this option to be valid, you should have added set of printing commands to the input file to print some (or all) of the results using the text editor. This is the old method of printing the results, which this tutorial will not focus on; as an alternative we will focus on the new method of viewing the results like the **Post Processing Mode**.
  - To **Go to Post Processing Mode**. With this option you will be directly taken to the **Post Processing Mode**, to view the results, which contains tables and graphical output in a very professional way.
  - To **Stay in Modeling Mode**. With this option you will stay at the **Modeling Mode**, and later on view the output file, or go to the **Post Processing Mode** at your convenience.

Second Case
- If STAAD.*Pro* find any type of error it will show the following dialog box:

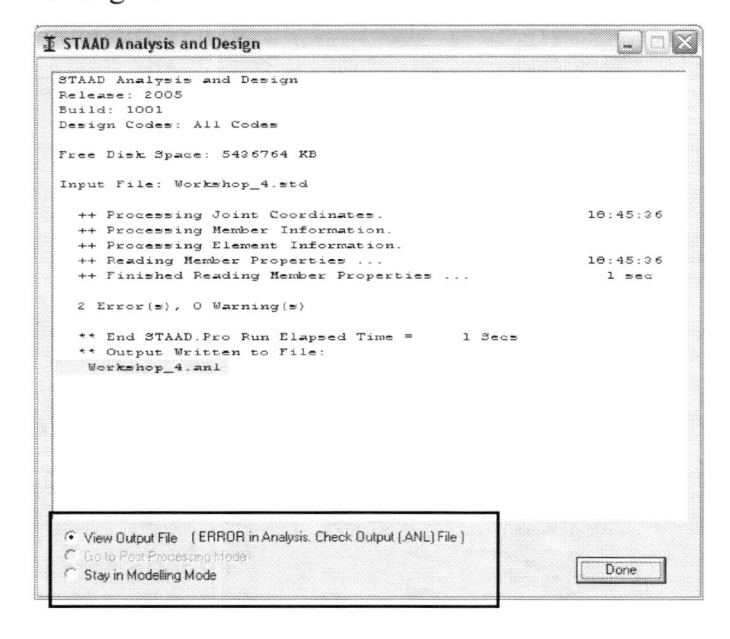

- In this dialog box, you can see that STAAD.*Pro* was still in reading input file, when it found that there is missing information which made the strcuture unstable.
- STAAD.*Pro* gave two choices to select from:
  - To **View Output File (ERROR in Analysis. Check Output (.ANL) File)**. This option will take you directly to the STAAD Output editor to view where your mistakes were, it will show something like the following:

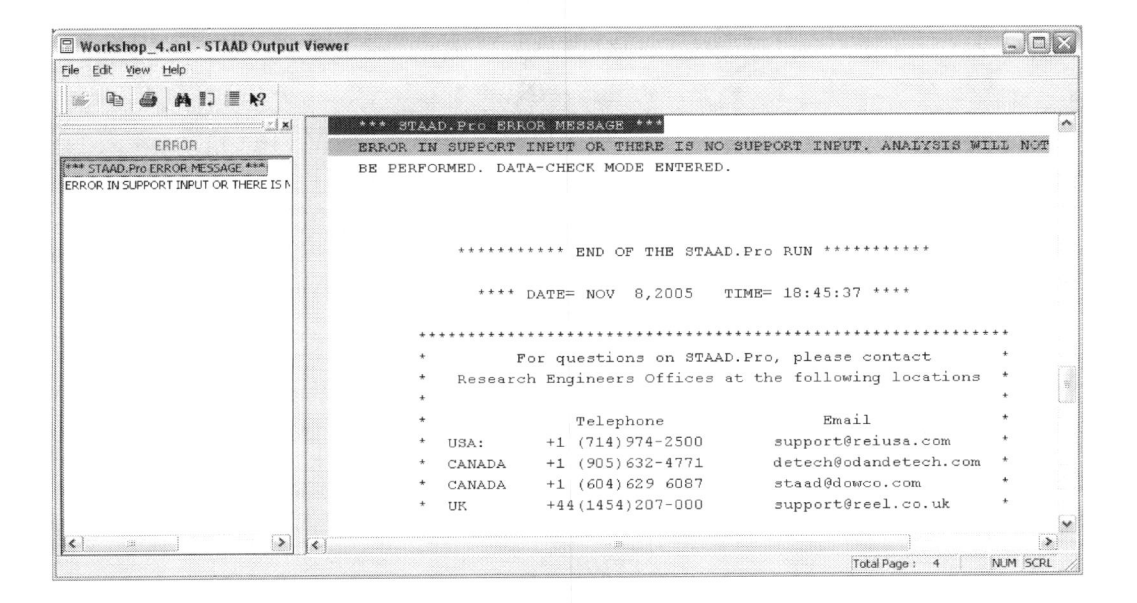

  - From the above message you will identify the error, hence go to the input file and correct it.
  - To **Stay in Modeling Mode**. With this option you will stay at the **Modeling Mode**, and later on view the output file, and find the error message.

Adding Analysis statement & Execution Command (Concrete & Steel Structure)

Workshop 5-A & 5-B

1. Open Small_Building file.
2. Add **Perform Analysis** Statement.
3. Select **Analyze/Run Analysis**.
4. If there are any errors try to go to the editor to identify the errors and correct them in your input file.
5. If needed, re-run the **Run Analysis** command.
6. If not, choose to stay in the **Modeling Mode**.
7. Save and close.

Module Review

1. STAAD.*Pro* Perform Analysis is:
    a. Taking into consideration the Displacement of Nodes.
    b. Taking into consideration the Stiffness Correction.
    c. Multi-iteration Analysis.
    d. None of the above.

2. P-Delta Analysis will ask the user for _____ and _____.

3. The maximum number of iterations in Non-Linear Analysis is 2:
    a. True
    b. False

4. ACI 318 is one of the codes recommends:
    a. P-Delta Analysis
    b. Non-Linear Analysis
    c. Time History Dynamic Analysis.
    d. Perform Analysis.

Module Review Answers

1. d
2. Number of iteration, and Convergence value.
3. b
4. a

Module 8:

Post Processing

This module contains:

- Post Processing Overview
- Node Results
- Beam Results
- Plate Results
- Creating Reports

Introduction

- Post Processing Mode is where you see your results, print them, or prepare a full report about selected parts.
- In our input file, we didn't specify what type of results we need because STAAD.*Pro* will produce them all for us.
- This way is better than old method used in STAAD-III, as user should specify in the input file what are the desired results, and STAAD.*Pro* will produce them only, nothing more.
- STAAD-III results were (low-quality) tables, which can fit the dot-matrix printers, with no graphics.
- STAAD.*Pro* results are graphical using colors in a very nice way, and handsome tables with fonts, and grid (which user can customize), and all can fit in the inkjet/laser printer with no additional hassle.
- Accordingly in this tutorial we will focus on the STAAD.*Pro* Post Processing with no mention to old ways of STAAD-III (which is still available in STAAD.*Pro*).

First Step

- There are three ways to go to Post Processing Mode:
  - From menus select **Mode/Post Processing**.
  - From **Mode** toolbar, select the **Post Processing** icon.
  - After you finish the **Run Analysis** command, the dialog box will give you an option to go directly to Post Processing Mode (refer to Module 7)

- Select any of the three methods, the following dialog box will be shown:

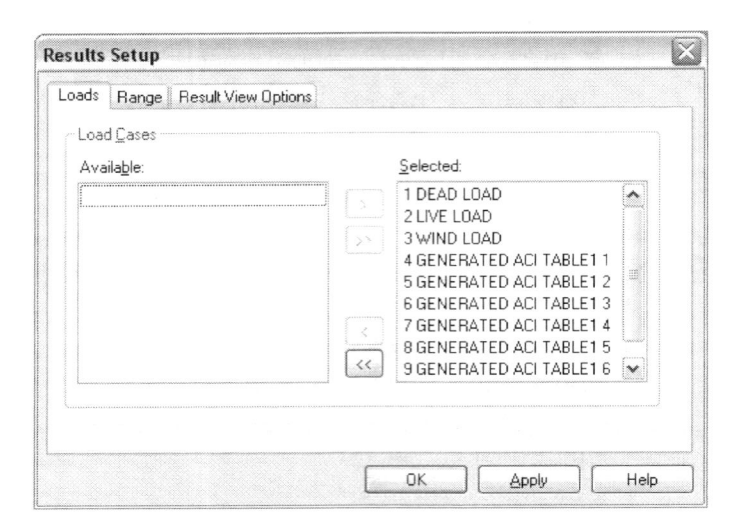

- User should specify the desired loads to see the results for. As you can see STAAD.*Pro* assumed that all loads are wanted, but this may generate huge tables of results, which is undesired, accordingly we will focus only on Combinations.

- Select the Primary Loads (click-and-drag, or use Ctrl key), and click the left single arrow, you will get the following:

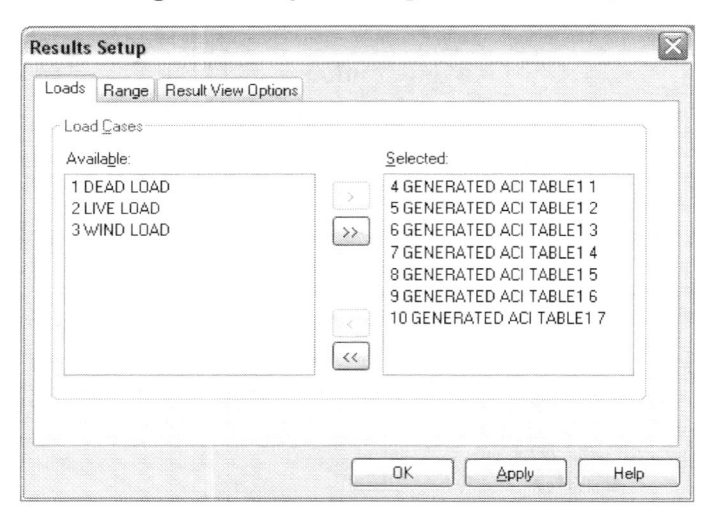

- Click **OK**, and you will be taken to the Post Processing Mode.

Module 8: Post Processing

Node Displacement

- The first Page will be opened to the user is the Node Displacement.
- You will see the Screen cut into three parts:
  - The Node Displacement table.
  - The Beam Relative Displacement Detail table.
  - The Node Displacement graphical shape.

Node Displacement table-*All* tab

- In this tab you will see for each Node, and for each load case selected by the user:
  - The movement of the Node in the X, Y, and Z direction.
  - The Resultant movement of the Node.
  - The Rotational of Node around X, Y, and Z measured in radians.
- See the below table as an example:

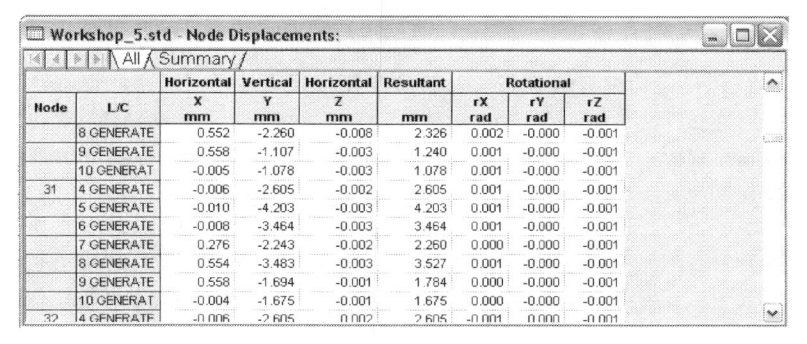

Node	L/C	Horizontal X mm	Vertical Y mm	Horizontal Z mm	Resultant mm	rX rad	rY rad	rZ rad
	8 GENERATE	0.552	-2.260	-0.008	2.326	0.002	-0.000	-0.001
	9 GENERATE	0.558	-1.107	-0.003	1.240	0.001	-0.000	-0.001
	10 GENERAT	-0.005	-1.078	-0.003	1.078	0.001	-0.000	-0.001
31	4 GENERATE	-0.006	-2.605	-0.002	2.605	0.001	-0.000	-0.001
	5 GENERATE	-0.010	-4.203	-0.003	4.203	0.001	-0.000	-0.001
	6 GENERATE	-0.008	-3.464	-0.003	3.464	0.001	-0.000	-0.001
	7 GENERATE	0.276	-2.243	-0.002	2.260	0.000	-0.000	-0.001
	8 GENERATE	0.554	-3.483	-0.003	3.527	0.001	-0.000	-0.001
	9 GENERATE	0.558	-1.694	-0.001	1.784	0.000	-0.000	-0.001
	10 GENERAT	-0.004	-1.675	-0.001	1.675	0.000	-0.000	-0.001
32	4 GENERATE	-0.006	-2.605	0.002	2.605	-0.001	0.000	-0.001

Node Displacement table-*Summary* tab

- In this tab you will see:
  - The Max and Min movement in the X, Y, and Z directions, and which load case caused it. (Max is the largest number in positive, and Min is the largest number in negative)
  - The max and min rotation around X, Y, and Z, and which load case caused it.
- See the below table as an example:

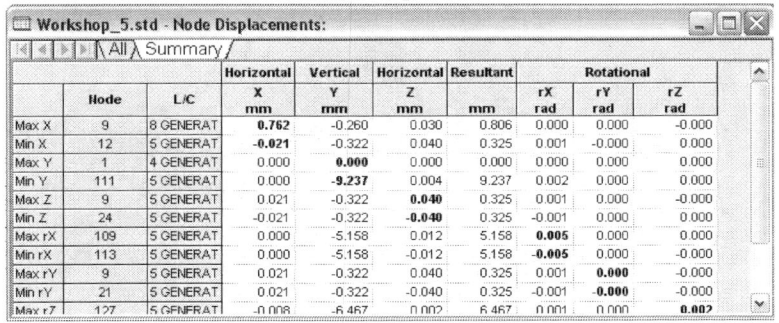

	Node	L/C	Horizontal X mm	Vertical Y mm	Horizontal Z mm	Resultant mm	rX rad	rY rad	rZ rad
Max X	9	8 GENERAT	**0.762**	-0.260	0.030	0.806	0.000	0.000	-0.000
Min X	12	5 GENERAT	**-0.021**	-0.322	0.040	0.325	0.001	-0.000	0.000
Max Y	1	4 GENERATE	0.000	**0.000**	0.000	0.000	0.000	0.000	0.000
Min Y	111	5 GENERAT	0.000	**-9.237**	0.004	9.237	0.002	0.000	0.000
Max Z	9	5 GENERAT	0.021	-0.322	**0.040**	0.325	0.001	0.000	-0.000
Min Z	24	5 GENERAT	-0.021	-0.322	**-0.040**	0.325	-0.001	0.000	0.000
Max rX	109	5 GENERAT	0.000	-5.158	0.012	5.158	**0.005**	0.000	0.000
Min rX	113	5 GENERAT	0.000	-5.158	-0.012	5.158	**-0.005**	0.000	-0.000
Max rY	9	5 GENERAT	0.021	-0.322	0.040	0.325	0.001	**0.000**	-0.000
Min rY	21	5 GENERAT	0.021	-0.322	-0.040	0.325	-0.001	**-0.000**	-0.000
Max rZ	127	5 GENERAT	-0.008	-6.467	0.002	6.467	0.001	0.000	**0.002**

8-5

Beam Relative Displacement Detail- *All* tab

- In this tab you will see for each Beam, and for each load case selected by the user, the movement in the X, Y, and Z direction of the Start Node, and the End Node, and three intermediate points inside the Beam (equally spaced), and finally the absolute Resultant movement.
- See the below table as an example:

Beam Relative Displacement Detail- *Max. Rel. Displacement* tab

- In this tab you will see for each Beam, and for each load case selected by the user, the maximum movement of the Beam and where it occur within the span of the Beam in the X, Y, and Z direction.
- Also you will see the absolute maximum movement in the three directions, and where it occurs within the span.
- Finally, you will see the Span/Max ratio. As an example, assume that a Beam with 4m long made 14.861 mm displacement, so, the Span/Max will be 4/0.014861 the result is 269.
- See the below table as an example:

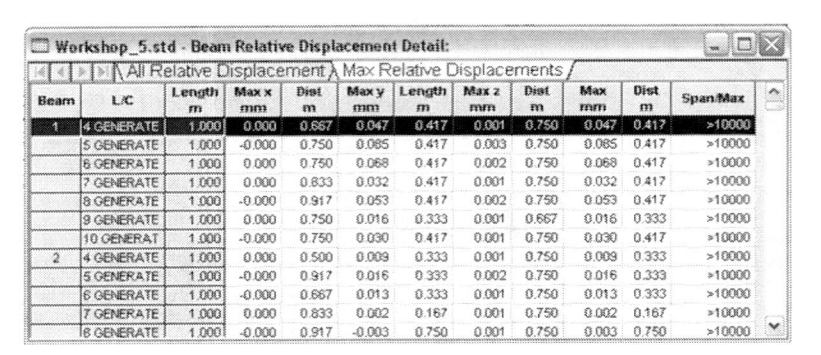

8-6

Module 8: Post Processing

Node Displacement Graphical Display

- You will see in this graphical display the Deflected shape of the structure.
- As a preparation do the following steps:
  - From the **View** toolbar, select the desired load case.

  - From the **Structure** toolbar, select **Change Graphical Display Unit** icon, to get the following dialog box:

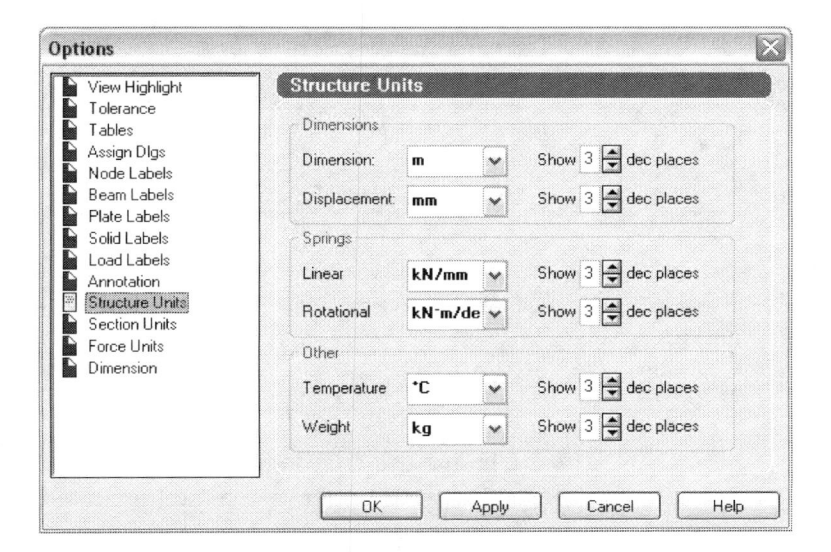

  - From the left list, select **Structure Units**.
  - Change the **Displacement** unit to be shown on the shape as either cm, or mm, or any convenient unit, also you can specify the decimal places.

8-7

- If the defelected shape is not clear, you need to fix the scale of the display, from the **Strcuture** toolbar, select **Scale** icon, you will see the following:

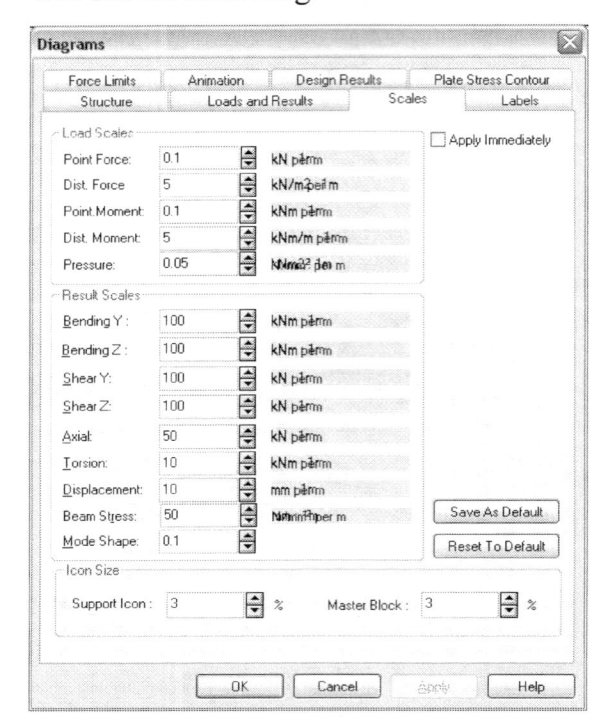

- Change the **Displacement** scale to be less number than the existing, in order to see the Deflected Shape more clearer.

■ You will see something similar to the following:

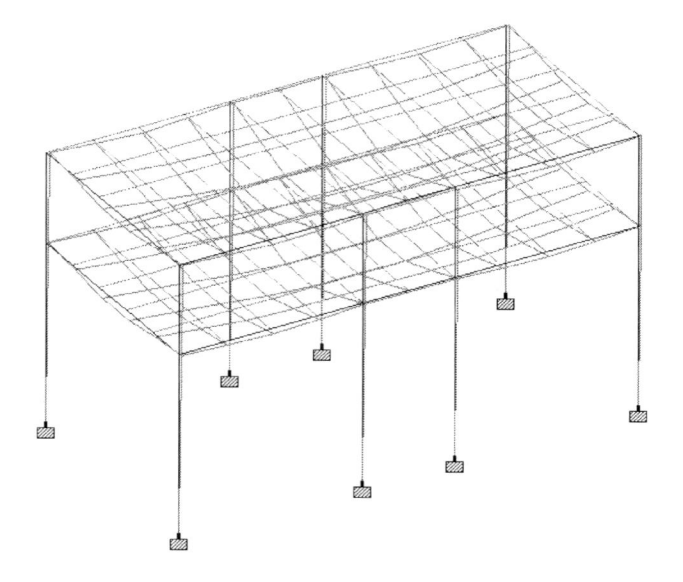

■

Module 8: Post Processing

- In order to see the values on the defelected shape, do the following steps:
  - From menus, select **Results/View Value**, and you will see the following dialog box:

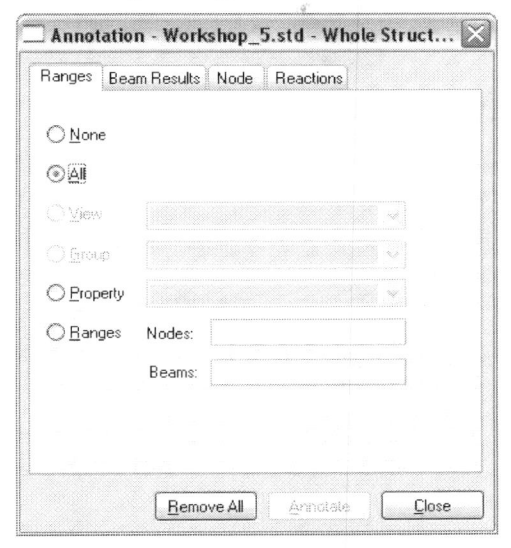

  - From the **Ranges** tab, click **All** (you can also select specific **Property** (Cross Section) or, you can specify **Ranges** of numbers for Nodes, and Beams)

  - Select the **Node** tab, you will see the following:

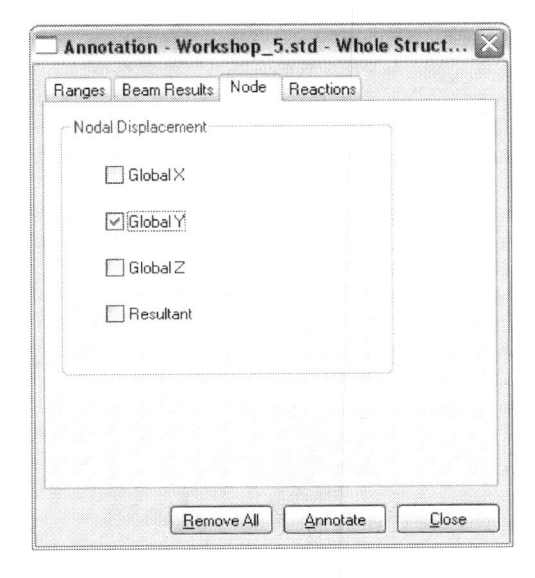

8-9

- Select the Nodal Displacement type, you have **Global X**, **Global Y**, **Global Z**, and finally the **Resultant**, click **Annotate**.

■ You will get something similar to:

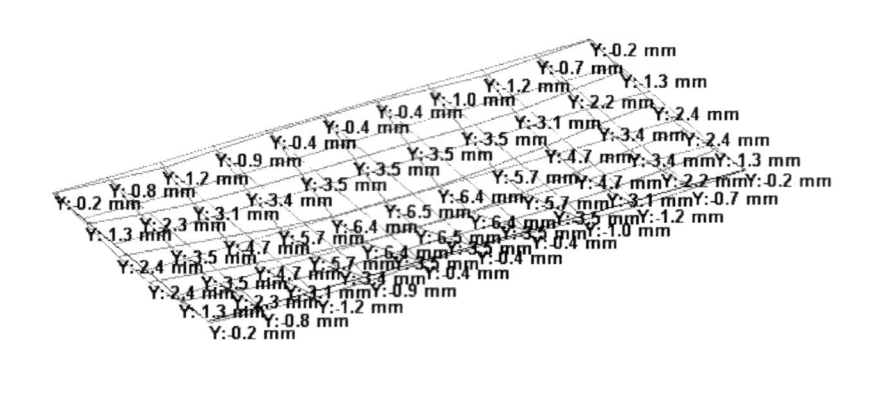

Node Reactions

■ You will see the Screen cut into two parts:
  - The Support Reaction table.
  - Graphical shape showing each support and its reactions.

Support Reaction table – *All* tab

■ In this tab you will see for each support, for each load case selected by the user, 6 reactions Fx, Fy, Fz, Mx, My, and Mz. See the below table as an example:

Node	L/C	Fx kN (Horizontal)	Fy kN (Vertical)	Fz kN (Horizontal)	Mx kNm	My kNm	Mz kNm
1	4 GENERATE	2.711	180.620	4.293	5.798	0.080	-3.653
	5 GENERATE	4.611	287.811	7.679	10.374	0.153	-6.210
	6 GENERATE	3.753	237.938	6.180	8.347	0.122	-5.056
	7 GENERATE	0.317	153.210	3.669	4.960	0.095	1.856
	8 GENERATE	-0.260	234.724	6.158	8.328	0.174	4.919
	9 GENERATE	-2.271	112.898	2.739	3.708	0.104	7.626
	10 GENERAT	1.743	116.113	2.760	3.727	0.052	-2.349
2	4 GENERATE	-1.465	222.088	4.446	5.922	0.009	1.915
	5 GENERATE	-2.503	377.367	8.033	10.701	0.014	3.278
	6 GENERATE	-2.035	307.240	6.450	8.592	0.012	2.659
	7 GENERATE	-3.269	187.942	3.807	5.072	0.007	6.251
	8 GENERATE	-6.061	302.401	6.443	8.582	0.011	11.879
	9 GENERATE	-4.968	137.932	2.851	3.798	0.005	10.450

Module 8: Post Processing

Support Reaction table – *Summary* tab
- In this tab you will see the maximum and minimum Fx, Fy, Fz, Mx, My, and Mz, and on which supports took place, and which of the selected load cases has caused them.
- See the below table as an example:

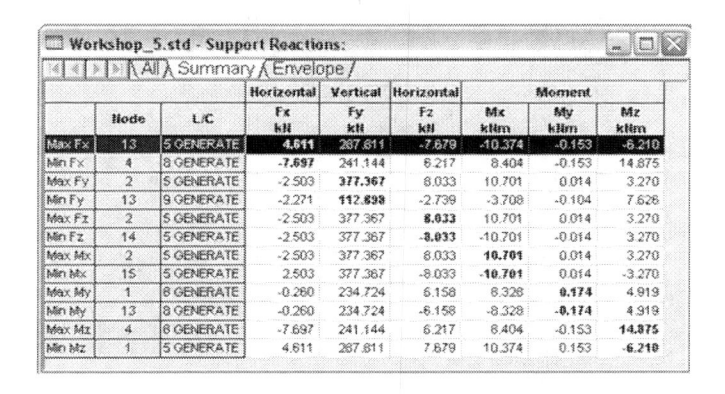

	Node	L/C	Horizontal Fx kN	Vertical Fy kN	Horizontal Fz kN	Mx kNm	My kNm	Mz kNm
Max Fx	13	5 GENERATE	4.611	287.811	-7.679	-10.374	-0.153	-6.210
Min Fx	4	8 GENERATE	-7.697	241.144	6.217	8.404	-0.153	14.875
Max Fy	2	5 GENERATE	-2.503	377.367	8.033	10.701	0.014	3.270
Min Fy	13	9 GENERATE	-2.271	112.698	-2.739	-3.708	-0.104	7.626
Max Fz	2	5 GENERATE	-2.503	377.367	8.033	10.701	0.014	3.270
Min Fz	14	5 GENERATE	-2.503	377.367	-8.033	-10.701	-0.014	3.270
Max Mx	2	5 GENERATE	-2.503	377.367	8.033	10.701	0.014	3.270
Min Mx	15	5 GENERATE	2.503	377.367	-8.033	-10.701	0.014	-3.270
Max My	1	8 GENERATE	-0.260	234.724	6.158	8.328	0.174	4.919
Min My	13	8 GENERATE	-0.260	234.724	-6.158	-8.328	-0.174	4.919
Max Mz	4	8 GENERATE	-7.697	241.144	6.217	8.404	-0.153	14.875
Min Mz	1	5 GENERATE	4.611	287.811	7.679	10.374	0.153	-6.210

Support Reaction table – *Envelope* tab
- In this tab you will find the maximum positive and the maximum negative Fx, Fy, Fz, Mx, My, and Mz on each support, and which of the selected load case caused them.
- See the below table as an example:

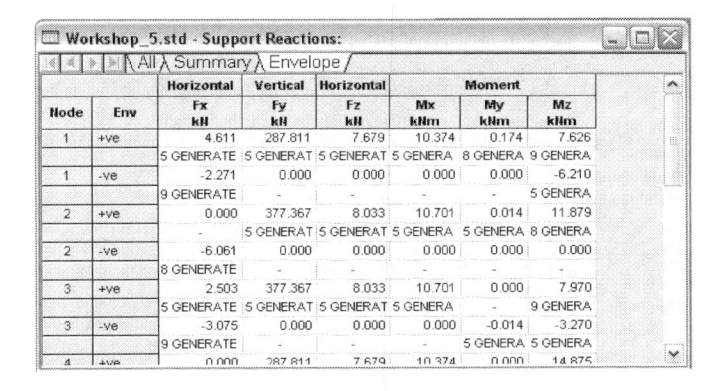

Node	Env	Fx kN	Fy kN	Fz kN	Mx kNm	My kNm	Mz kNm
1	+ve	4.611	287.811	7.679	10.374	0.174	7.626
		5 GENERATE	5 GENERAT	5 GENERAT	5 GENERA	8 GENERA	9 GENERA
1	-ve	-2.271	0.000	0.000	0.000	0.000	-6.210
		9 GENERATE	-	-	-	-	5 GENERA
2	+ve	0.000	377.367	8.033	10.701	0.014	11.879
		-	5 GENERAT	5 GENERAT	5 GENERA	5 GENERA	8 GENERA
2	-ve	-6.061	0.000	0.000	0.000	0.000	0.000
		8 GENERATE	-	-	-	-	-
3	+ve	2.503	377.367	8.033	10.701	0.000	7.970
		5 GENERATE	5 GENERAT	5 GENERAT	5 GENERA	-	9 GENERA
3	-ve	-3.075	0.000	0.000	0.000	-0.014	-3.270
		9 GENERATE	-	-	-	5 GENERA	5 GENERA
4	+ve	0.000	287.811	7.679	10.374	0.000	14.875

Support Reaction Graphical Display
- In this display the user will see the 6 reactions of each support.
- As a preparation select the desired load case from the **View** toolbar

8-11

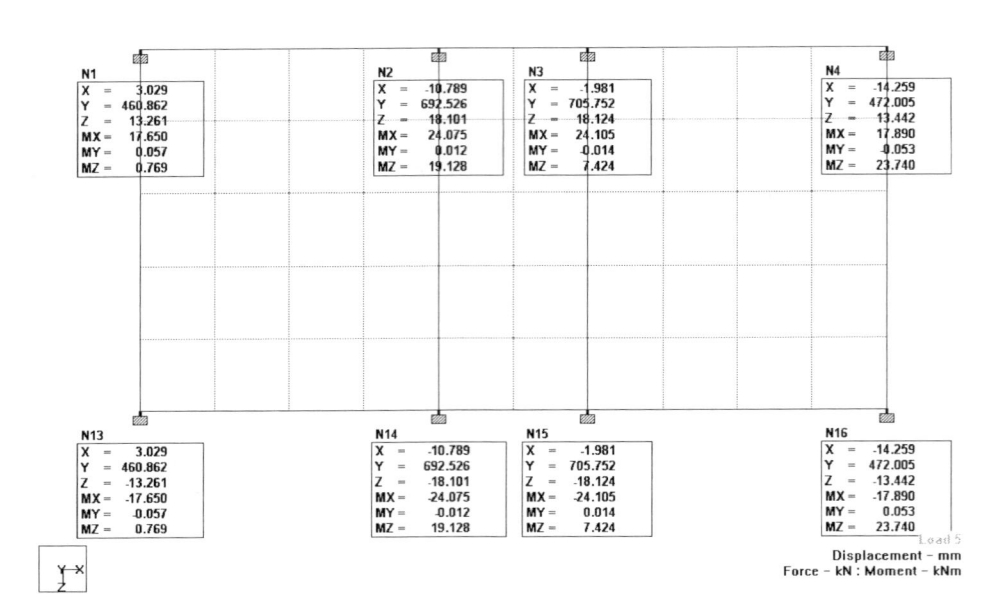

- You will see something like that (this is a top view):

[Diagram showing top view with nodes N1, N2, N3, N4 on top row and N13, N14, N15, N16 on bottom row, each with X, Y, Z, MX, MY, MZ values]

N1: X = 3.029, Y = 460.862, Z = 13.261, MX = 17.650, MY = 0.057, MZ = 0.769
N2: X = -10.789, Y = 692.526, Z = 18.101, MX = 24.075, MY = 0.012, MZ = 19.128
N3: X = -1.981, Y = 705.752, Z = 18.124, MX = 24.105, MY = -0.014, MZ = 7.424
N4: X = -14.259, Y = 472.005, Z = 13.442, MX = 17.890, MY = -0.053, MZ = 23.740

N13: X = 3.029, Y = 460.862, Z = -13.261, MX = -17.650, MY = -0.057, MZ = 0.769
N14: X = -10.789, Y = 692.526, Z = -18.101, MX = -24.075, MY = -0.012, MZ = 19.128
N15: X = -1.981, Y = 705.752, Z = -18.124, MX = -24.105, MY = 0.014, MZ = 7.424
N16: X = -14.259, Y = 472.005, Z = -13.442, MX = -17.890, MY = 0.053, MZ = 23.740

Load 5
Displacement – mm
Force – kN : Moment – kNm

Beam Forces

- You will see the Screen cut into three parts:
  - The Beam End Forces table.
  - The Beam Force Detail table.
  - Graphical Display showing (by default) the Bending Moment Diagram using the current load case.

Beam End Forces table – *All* tab

- In this tab you will see for each Beam, for each load case selected by the user, for each Node at the ends of the Beam the 6 reactions Fx, Fy, Fz, Mx, My, and Mz.
- See the below table as an example:

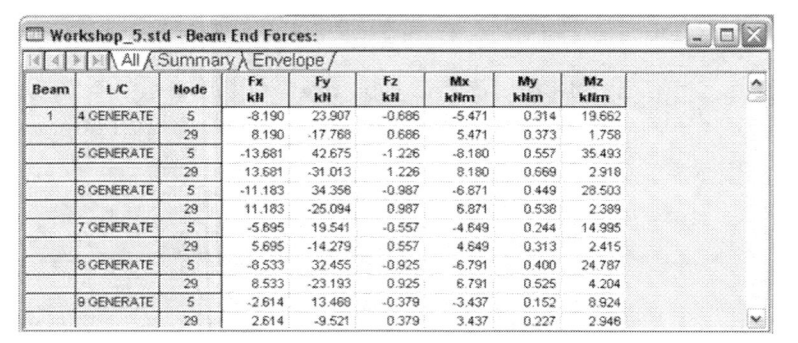

Beam	L/C	Node	Fx kN	Fy kN	Fz kN	Mx kNm	My kNm	Mz kNm
1	4 GENERATE	5	-8.190	23.907	-0.686	-5.471	0.314	19.662
		29	8.190	-17.768	0.686	5.471	0.373	1.758
	5 GENERATE	5	-13.681	42.675	-1.226	-8.180	0.557	35.493
		29	13.681	-31.013	1.226	8.180	0.669	2.918
	6 GENERATE	5	-11.183	34.356	-0.987	-6.871	0.449	28.503
		29	11.183	-25.094	0.987	6.871	0.538	2.389
	7 GENERATE	5	-5.695	19.541	-0.557	-4.649	0.244	14.995
		29	5.695	-14.279	0.557	4.649	0.313	2.415
	8 GENERATE	5	-8.533	32.455	-0.925	-6.791	0.400	24.787
		29	8.533	-23.193	0.925	6.791	0.525	4.204
	9 GENERATE	5	-2.614	13.468	-0.379	-3.437	0.152	8.924
		29	2.614	-9.521	0.379	3.437	0.227	2.946

Beam End Forces table – *Summary* tab

- In this tab you will see the maximum and minimum Fx, Fy, Fz, Mx, My, and Mz, and on which Node of each Beam, and which of the selected load cases has caused them.
- See the below table as an example:

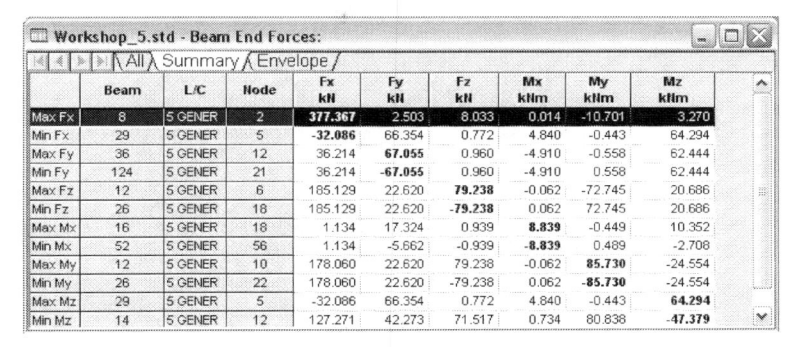

Beam End Forces table – *Envelope* tab

- In this tab you will find the maximum positive and the maximum negative Fx, Fy, Fz, Mx, My, and Mz on each Node of each Beam, and which of the selected load case caused them. See the below table as an example:

Beam Force Detail table – *All* tab

- Same as Beam End Forces – All tab, but instead it shows the 3 forces and 3 moments for the beginning of the Beam (Dist = 0.0) and the ending of the Beam (Dist = 1.0), and for 3 intermidiate distances (1/4 of the length, 1/2 of the of the length, and 3/4 of the length).

Beam Force Detail table – *Max Axial Forces* tab

- This tab will show the maximum positive and negative axial forces (Fx) for each Beam, for each load case, and where it occur:

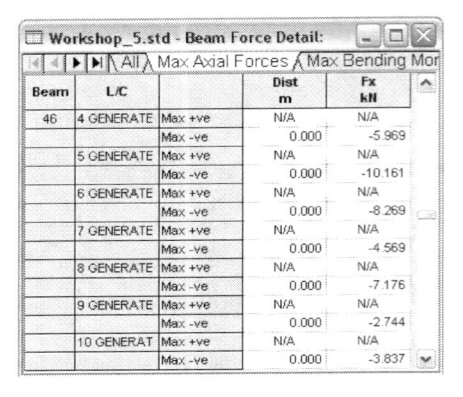

Beam Force Detail table – *Max Bending Moments* tab

- This tab will show the maximum positive and negative Bending Moments (Mz, and My) for each Beam, for each load case, and where it occur:

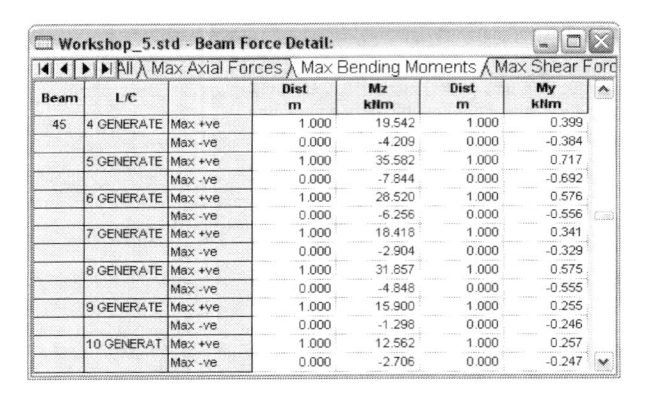

Beam Force Detail table – *Max Shear Forces* tab

- This tab will show the maximum positive and negative Shear Forces (Fy, and Fz) for each Beam, for each load case, and where it occur:

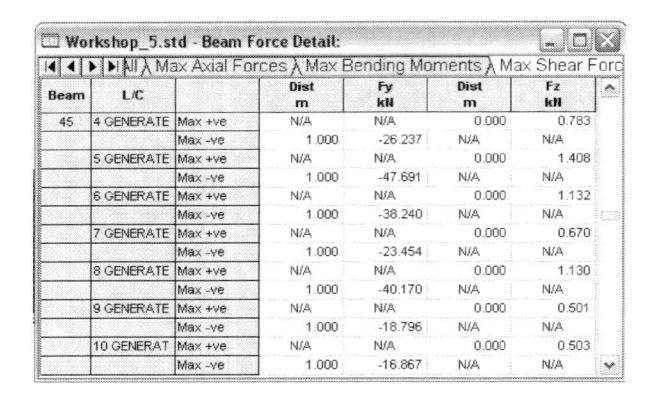

Module 8: Post Processing

Beam End Forces Graphical Display

- By default the user will see the Bending Moment Diagram for the current load case.
- To change the current load case, go to the **View** toolbar, and change it (you can select to view the Envelope):

- From the **Results** toolbar, you can switch ON/OFF the results you would like to view:

  - To view the Axial Load Fx
  - To view the Shear in Y direction Fy
  - To view the Shear in Z direction Fz
  - To view Moment around X (Torsion) Mx
  - To view Moment around Y My
  - To view Bending Moment around Z Mz

- You can change the scale of Bending, Axial, Shear, etc., just like we discussed it in the Node Displacement Graphical Display.
- Also, you can Show/Hide the values of the different results on the Beam End Forces Graphical Display, using from the menus **Results/View Value**, then Select the **Beam Results** tab, you will see the following dialog box:

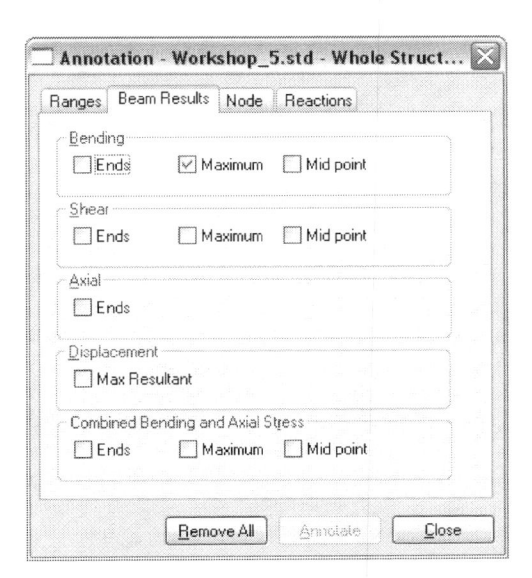

8-15

- Select the type of results you would like to show, and click **Annotate**. You will see something like the following:

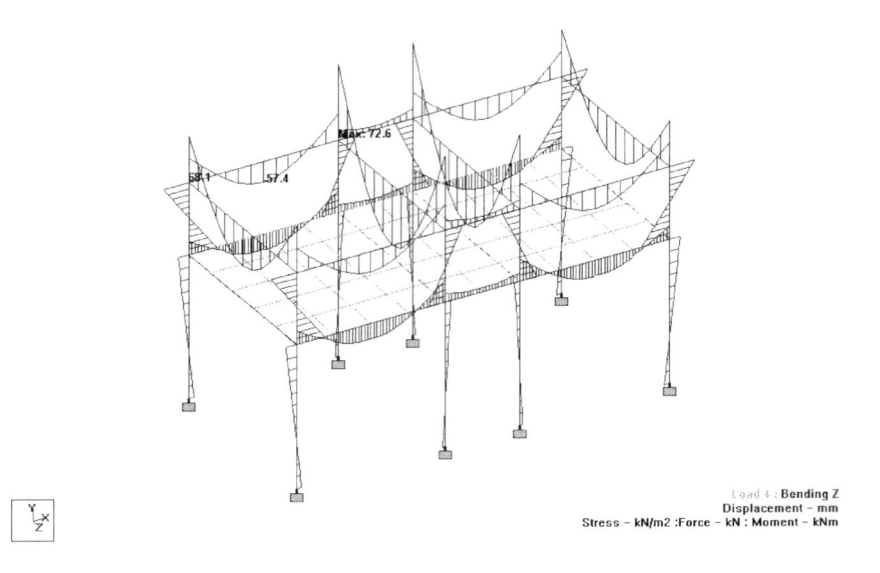

Beam Stresses

- You will see the Screen cut into five parts:
  - The Beam Combined Axial and Bending Stresses table.
  - Graphical Display showing the stresses for the current load case.
  - 3D Beam Stress Contour Display.
  - Dialog box titled **Select Section Plane** with slider, allows user to specify section plane along the length to know the stress values.

Beam Combined Axial and Bending Stresses table- *All* **tab**

- In this tab you will see for each Beam, for each load case selected by the user, and at the Start, End, and three intermediate points the values of the Stresses at the four corners of the Beam cross section. Then table lists the Max Compressive Stress and the Max Tensile Stress. See the below table as an example:

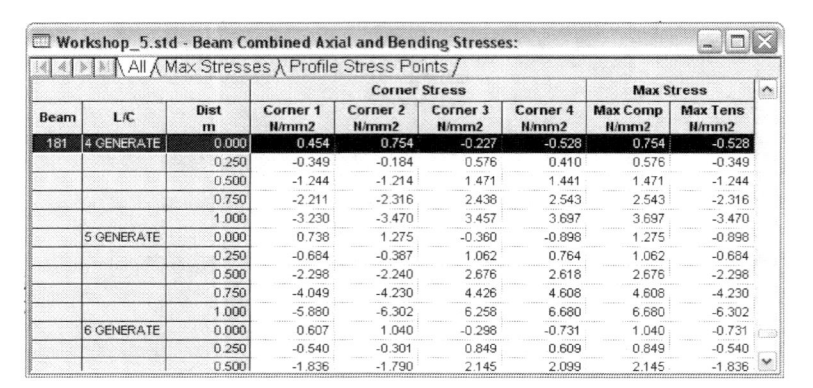

Beam	L/C	Dist m	Corner 1 N/mm2	Corner 2 N/mm2	Corner 3 N/mm2	Corner 4 N/mm2	Max Comp N/mm2	Max Tens N/mm2
181	4 GENERATE	0.000	0.454	0.754	-0.227	-0.528	0.754	-0.528
		0.250	-0.349	-0.184	0.576	0.410	0.576	-0.349
		0.500	-1.244	-1.214	1.471	1.441	1.471	-1.244
		0.750	-2.211	-2.316	2.438	2.543	2.543	-2.316
		1.000	-3.230	-3.470	3.457	3.697	3.697	-3.470
	5 GENERATE	0.000	0.738	1.275	-0.360	-0.898	1.275	-0.898
		0.250	-0.684	-0.387	1.062	0.764	1.062	-0.684
		0.500	-2.298	-2.240	2.676	2.618	2.676	-2.298
		0.750	-4.049	-4.230	4.426	4.608	4.608	-4.230
		1.000	-5.880	-6.302	6.258	6.680	6.680	-6.302
	6 GENERATE	0.000	0.607	1.040	-0.298	-0.731	1.040	-0.731
		0.250	-0.540	-0.301	0.849	0.609	0.849	-0.540
		0.500	-1.836	-1.790	2.145	2.099	2.145	-1.836

Module 8: Post Processing

Beam Combined Axial and Bending Stresses table- *Max Stresses* tab

- In this tab you will see for each Beam, for each load case selected by the user; the length of each Beam, the Maximum Compressive Stress, where along the length, and in which of the four corners occurred, the Maximum Tensile Stress, where along the length, and in which of the four corners occurred. See the below table as an example:

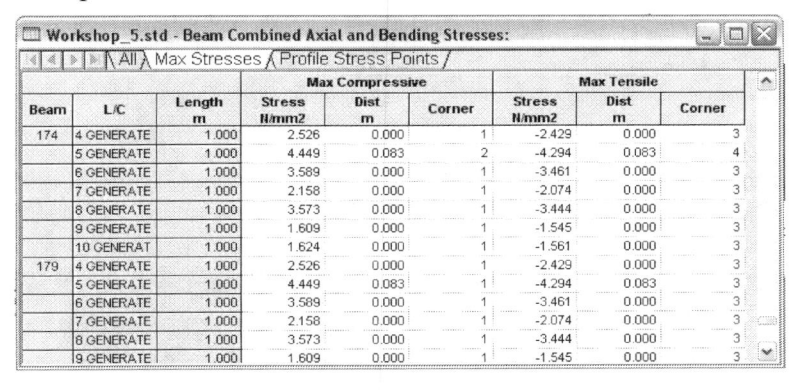

Beam	L/C	Length m	Max Compressive Stress N/mm2	Dist m	Corner	Max Tensile Stress N/mm2	Dist m	Corner
174	4 GENERATE	1.000	2.526	0.000	1	-2.429	0.000	3
	5 GENERATE	1.000	4.449	0.083	2	-4.294	0.083	4
	6 GENERATE	1.000	3.589	0.000	1	-3.461	0.000	3
	7 GENERATE	1.000	2.158	0.000	1	-2.074	0.000	3
	8 GENERATE	1.000	3.573	0.000	1	-3.444	0.000	3
	9 GENERATE	1.000	1.609	0.000	1	-1.545	0.000	3
	10 GENERAT	1.000	1.624	0.000	1	-1.561	0.000	3
179	4 GENERATE	1.000	2.526	0.000	1	-2.429	0.000	3
	5 GENERATE	1.000	4.449	0.083	1	-4.294	0.083	3
	6 GENERATE	1.000	3.589	0.000	1	-3.461	0.000	3
	7 GENERATE	1.000	2.158	0.000	1	-2.074	0.000	3
	8 GENERATE	1.000	3.573	0.000	1	-3.444	0.000	3
	9 GENERATE	1.000	1.609	0.000	1	-1.545	0.000	3

Beam Combined Axial and Bending Stresses Graphical Display

- In this Display you will see the Compression and Tension Diagrams (the red is Compression, and Blue is Tension)
- Change the current load case, go to the **View** toolbar, and change it:

- Use the Scale function to view the right Display of the Stresses.
- Also, you can use the **Results/View Value** as well.
- You will get something like the following:

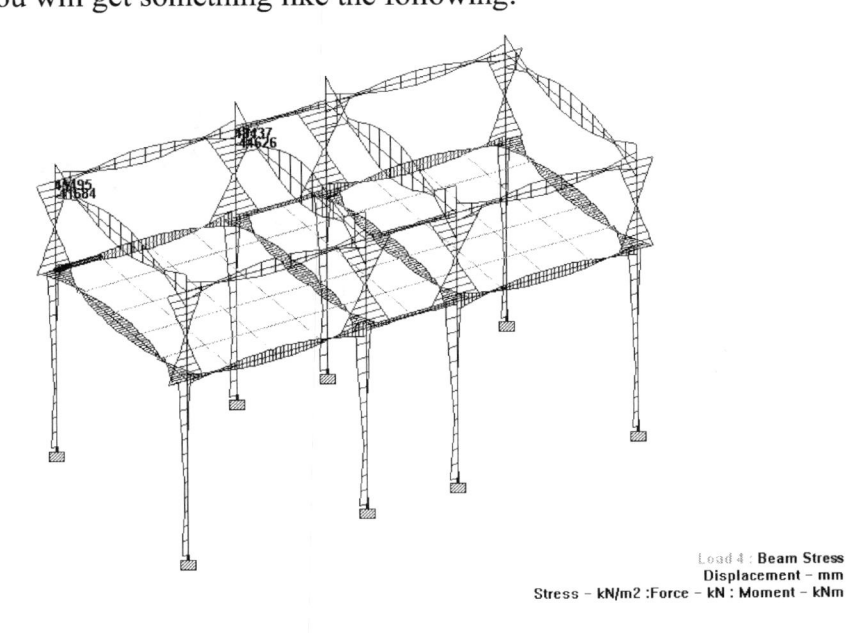

Load 4 : Beam Stress
Displacement – mm
Stress – kN/m2 :Force – kN : Moment – kNm

3D Beam Stress Contour

- In the upper part of the screen, you will find two windows showing:
  - At the left window is the full length of the selected Beam showing the stress contour in colors. Along the length you will find a cutting-plane, which can be moved through the slider in the **Select Section Plane** dialog box, as shown below:

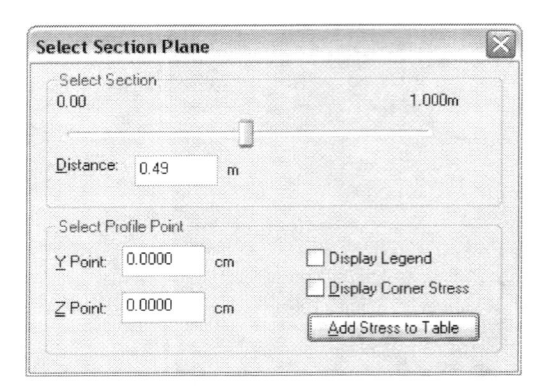

- Move the slider or type the **Distance** in which you want to view the stress at.
- Select to **Display Legend**, and to **Display Corner Stress**.
- In the right-side window of the upper part you will see something similar to the following:

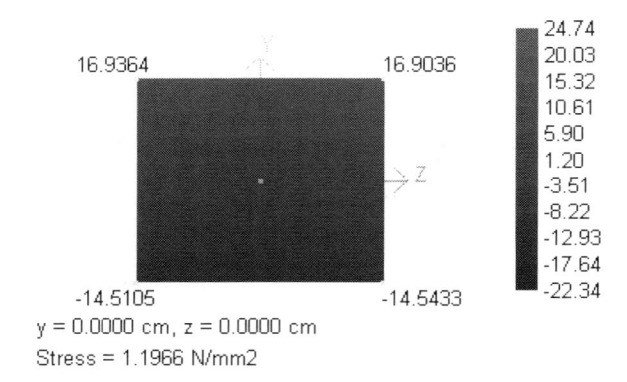

8-18

Beam Graphs

- This is a Beam-per-Beam Graphical result. When you click the sub-page of the Graphs something similar to the following will be displayed:

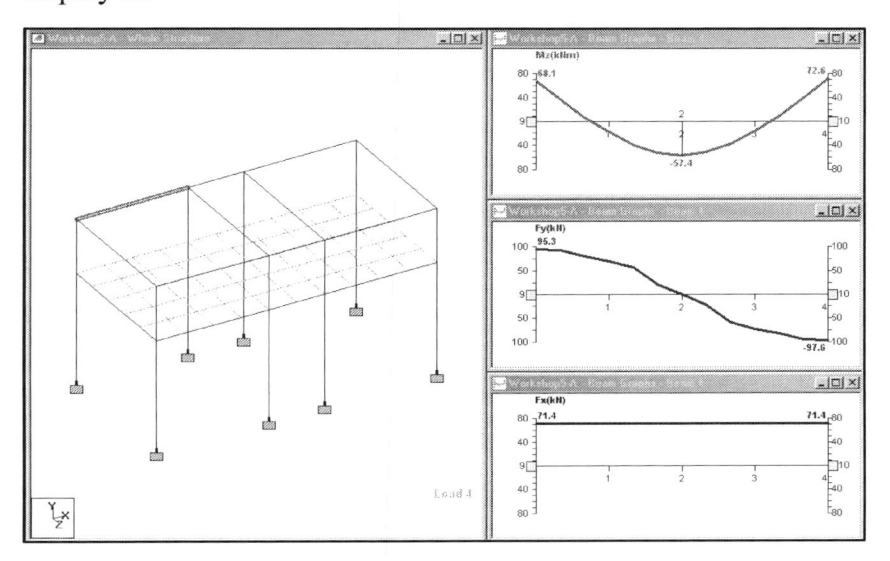

- From the **View** toolbar select the desired load case.
- It will display for the selected Beam, for the selected Load case:
  - Bending Moment Diagram (Moment around z-z).
  - Shear Diagram (Shear in y-y direction)
  - Axial Load Diagram.

Note
- You can change the three preset diagrams, by right-clicking on one of the diagrams and selecting **Diagrams** from the shortcut menu, the following dialog boxes will appear, Select the proper diagram you want, and click **OK**.

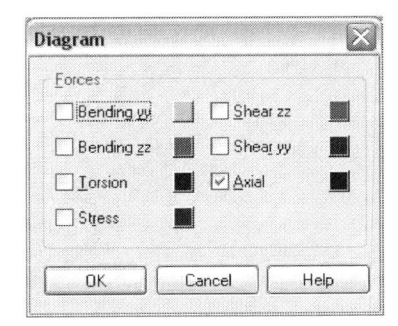

Plate Contour

- To view this result, your model should have plates.
- Once you go to this sub-page, the following dialog box will appear:

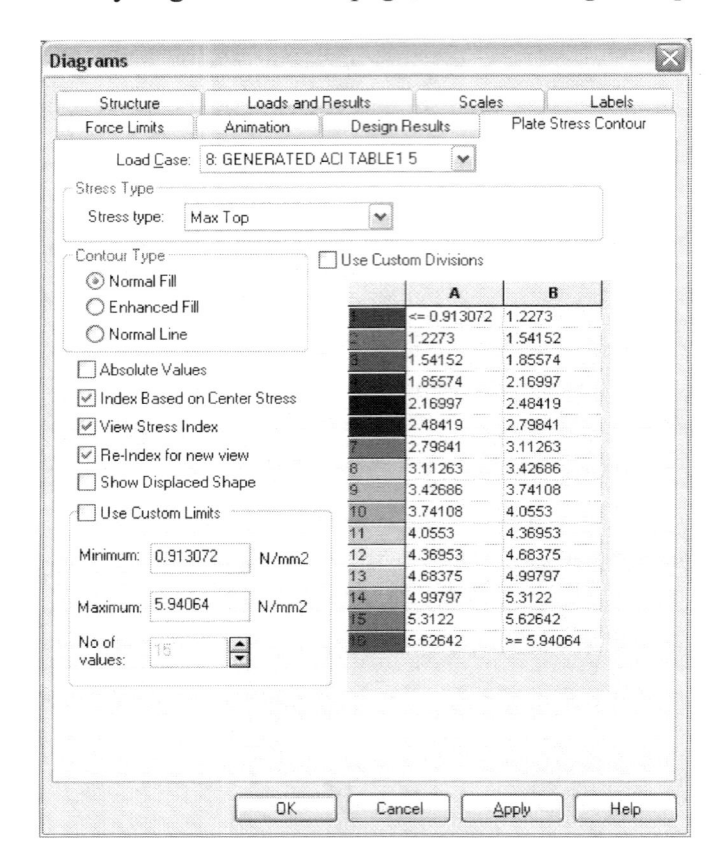

Module 8: Post Processing

- This dialog box will allow the user to control the Plate Stess Contour by changing the following:
  - The desired **Load Case** user likes to display the Stresses for.
  - **Stress Type**, you can choose one of the 19 available Stresses Types (please refer to the Technical Reference manual section 1.6.1 which discusses these types in details). Once you do, the Minimum, and Maximum value of this type will be displayed at the lower left corner of the dialog box.
  - **Contour Type**, select between **Normal Fill**, **Enhanced Fill**, or **Normal Lines**. **Nomal Fill** mode means the contour will be drawn using 5 points; the 4 edges, and the center of the plate. **Enhanced Fill** will use 5 points plus the mid point of each edge using interpolation. The **No. of values** will control the number of values to be displayed in the legend (max number is 15). **Normal Lines** means lines will be shown instead of fill.
  - Switch ON/OFF the **Absolute Values** checkbox. If this is ON that means all the Stesses value will be turned to postive numbers, and hence the comparison will be according to value and not to the sign.
  - Select to show the index (of colors) based on **Center Stress** or not?
  - Select **View Stress Index** or not on the left side of the screen.
  - Select to **Re-index** for any **new view**.
  - Select to **Show Displaced Shape** or, not?

8-21

STAAD.Pro 2005 Tutorial

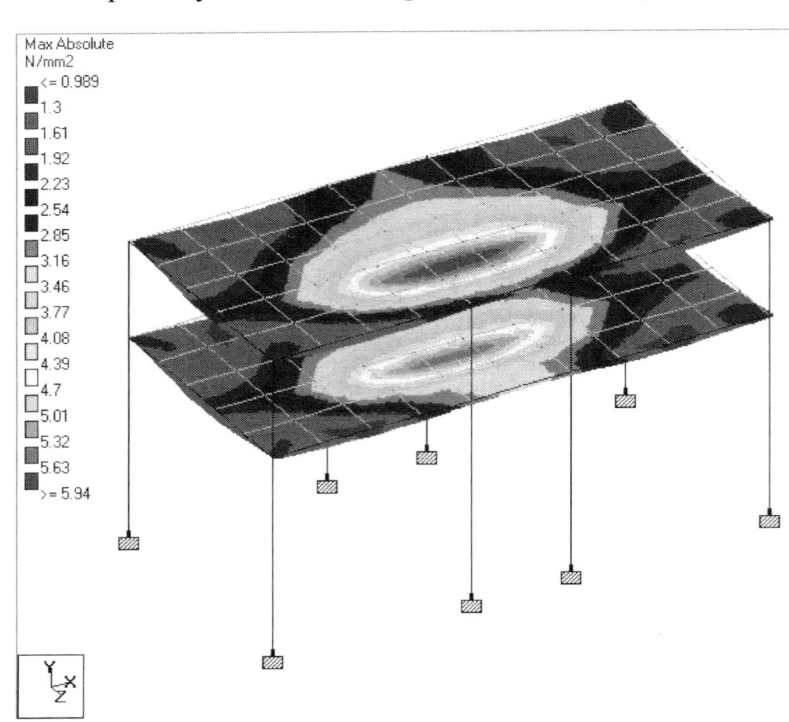

- The output may look something like the following:

Plate Centre Stresses Tables & Plate Corner Stress Tables

- After you read section 1.6.1 of the Technical Reference manual, and understand the types of stresses, take a look on these tables to view for each plate for each load case selected by the user the different types of stresses.
- One of the tables may look like the following:

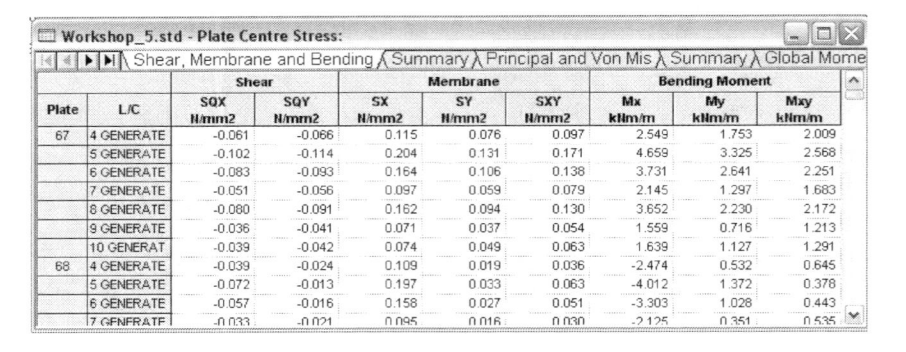

Plate	L/C	Shear SQX N/mm2	Shear SQY N/mm2	Membrane SX N/mm2	Membrane SY N/mm2	Membrane SXY N/mm2	Bending Moment Mx kNm/m	Bending Moment My kNm/m	Bending Moment Mxy kNm/m
67	4 GENERATE	-0.061	-0.066	0.115	0.076	0.097	2.549	1.753	2.009
	5 GENERATE	-0.102	-0.114	0.204	0.131	0.171	4.659	3.325	2.568
	6 GENERATE	-0.083	-0.093	0.164	0.106	0.138	3.731	2.641	2.251
	7 GENERATE	-0.051	-0.056	0.097	0.059	0.079	2.145	1.297	1.683
	8 GENERATE	-0.080	-0.091	0.162	0.094	0.130	3.652	2.230	2.172
	9 GENERATE	-0.036	-0.041	0.071	0.037	0.054	1.559	0.716	1.213
	10 GENERAT	-0.039	-0.042	0.074	0.049	0.063	1.639	1.127	1.291
68	4 GENERATE	-0.039	-0.024	0.109	0.019	0.036	-2.474	0.532	0.645
	5 GENERATE	-0.072	-0.013	0.197	0.033	0.063	-4.012	1.372	0.378
	6 GENERATE	-0.057	-0.016	0.158	0.027	0.051	-3.303	1.028	0.443
	7 GENERATE	-0.033	-0.021	0.095	0.016	0.030	-2.125	0.351	0.535

8-22

Plate Results Along Line

- In this tab you can define a line along a series of plates to view the stresses and total forces along this line.
- When you select this tab, the screen will look like the following illustration below:

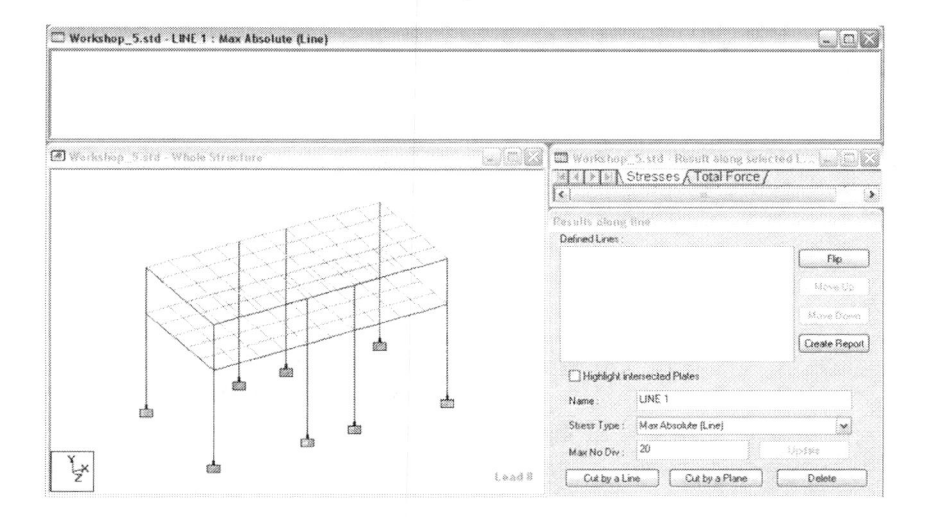

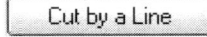

- From the **Results along line** dialog box, click **Cut by a Line** button.
- If you move the cursor over the structure you will notice that the cursor became a cross-hair.
- To define the line, click two points along the desired group of plates.
- A third point (which will be perpendicular on the defined line) will be asked as the direction of local Y).

- Once you are done, the screen will look something like the following:

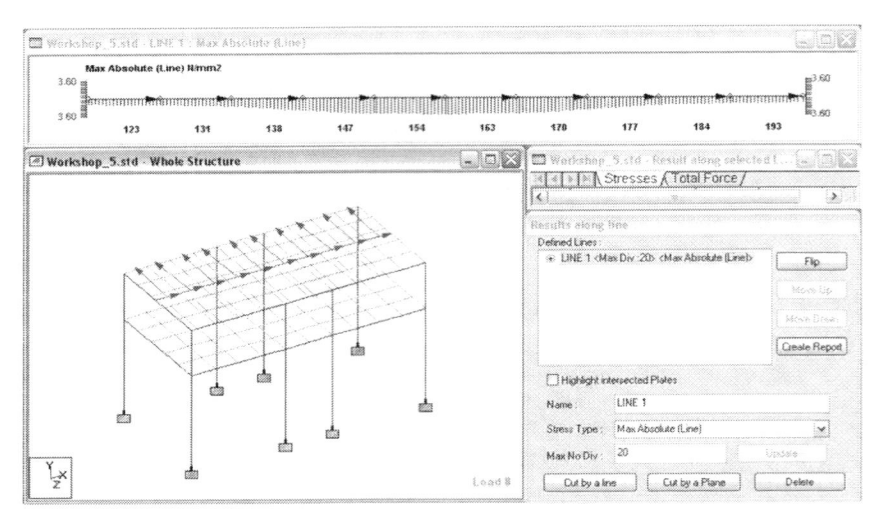

- The upper portion of the screen will show you the stress diagram along the line the user defined.
- Increase the **Max No. Div** from the **Results along line** dialog box, to view the upper portion of the screen in more details (the defualt is 2), then click **Update** button.
- Also, you can see the results as a table from the **Results along selected line** table, like the one below:

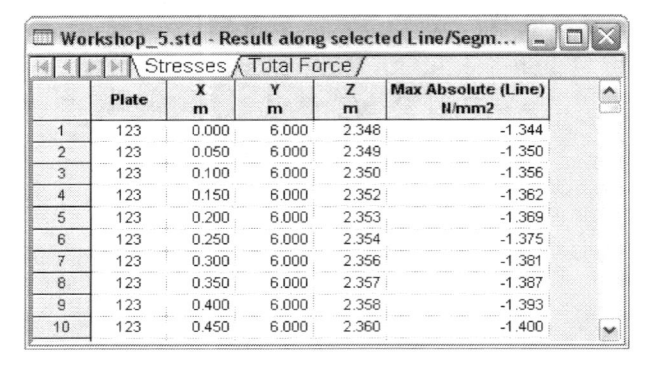

Module 8: Post Processing

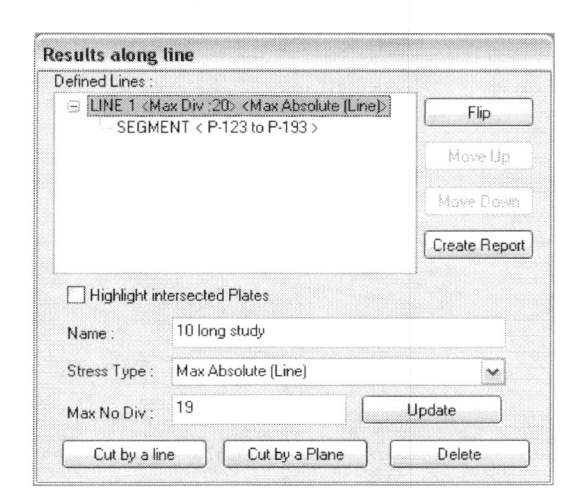

- In the **Results along line** dialog box like the one below, you will see the following:

    ![Results along line dialog box showing Defined Lines with LINE 1 <Max Div :20> <Max Absolute (Line)> and SEGMENT < P-123 to P-193 >. Buttons: Flip, Move Up, Move Down, Create Report. Checkbox: Highlight intersected Plates. Name: 10 long study. Stress Type: Max Absolute (Line). Max No Div: 19. Buttons: Update, Cut by a line, Cut by a Plane, Delete.]

    - By default, STAAD.*Pro* will call your line LINE1, but you can rename it as you wish.

    - Select the type of stress you would like to view.

    - Specify the **Max No Div** (as discussed earlier).

    - If you made any changes, click **Update** button to make these changes current.

    - In the Defined Lines part, you will see the name of the Line, the number of divisions, and the type of stress you select.

    - **Segment** (as descendant from the line) will show the number of the plate, which the line starts with, and the number of the plate which the line ends with.

    - In our example the Segment is showing P-123 to P-193, which means starting with plate # 123 and ending with plate # 193.

8-25

Animation

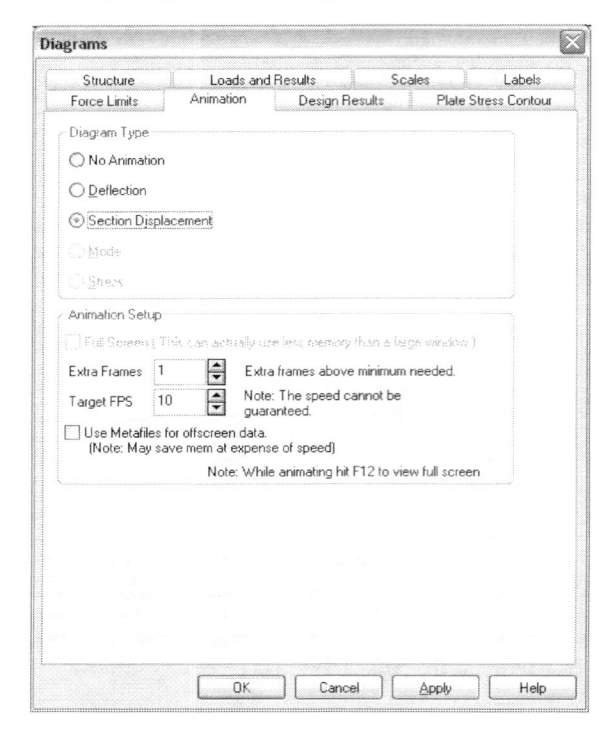

- Once you go to this Page, the following dialog box will appear:

- Select the **Diagram Type** from the following list:
  - **No Animation** (default option)
  - **Deflection** (means the movement of the Nodes)
  - **Section Displacement** (means the movement of the Nodes, and Deflection of the Beams)
- Set the **Animation setup** through changing the following:
  - **Extra Frames**, if you want more frames than the needed number in order to enhance the animation.
  - **Target FPS** (Frame Per Second); change this value to increase/decrease the speed of the animation.
- Click **OK**, to see the animation. Press F12 to see it in full screen.

Note
- The scale of Node Displacement affects the clarity of the animation.

Reports

- This is the replacement of all old fashion reports of STAAD-III.
- It should produce elegant-good-looking reports, which take advantage of the colors of ink-jet printers, and to include the graphical and tables in the same report.
- Once you go to this sub-page, the **Report Setup** dialog box will appear automatically:

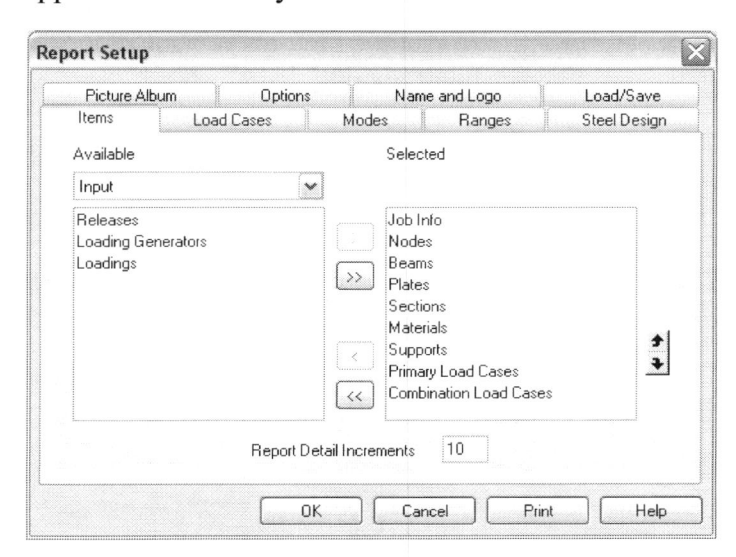

- Other ways to start the same command are:
  - From the **Print** toolbar, click **Report Setup** icon.
  - From menus, choose **File/Report Setup**.

Items tab
- In this tab you will select what are the contents of the report from the input data, output results, and pictures.
- From the **Available** pop-up list, select the desired part.

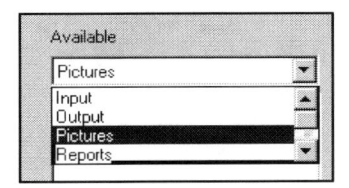

- From the **Input**, you will have a list of the Input data at the left. Select the input data you want to include it in your report. Then click the one arrow button (by default the Job Info is included in the report).
- From the **Output**, the same thing applies. You will see a list at the left. Select the output tables you want to include in your report; which they are identical to the tables we covered in this module. Then click the one arrow button.
- In order to include pictures in your report, you should capture the pictures before hand.
- While you are reviewing the results in the **Post Processing** mode you could use one of the following methods:

  - From the **Print** toolbar, click **Take Picture** icon.
  - From menus choose **Edit/Take Picture**.
  - The following dialog box will appear:

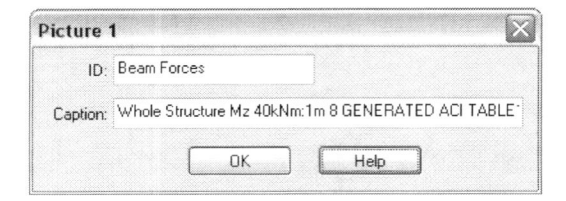

- Type in the **ID** field, which is the name of the picture. The **Caption** is generated automatically from the status of the results viewing on the screen.
- If you create several pictures, and you selected the **Pictures** option in the **Available** pop-up list, you will find these pictures listed for you. Select the desired pictures and click one arrow button.

- After you select the **Input**, **Output**, and **Pictures**, you may want to re-arrange the sequence of the data. You can use the two arrows at the right side of the dialog box. The arrow pointing up will take any item one step up in the list, hence this data will appear in the starting pages. The arrow pointing down will take any item one step down in the list, hence this data will appear in the ending pages

Load Cases tab
- By default the output results will be shown for all load cases (Primary & Combination).
- In this dialog box you can specify the load cases to be included in your report.

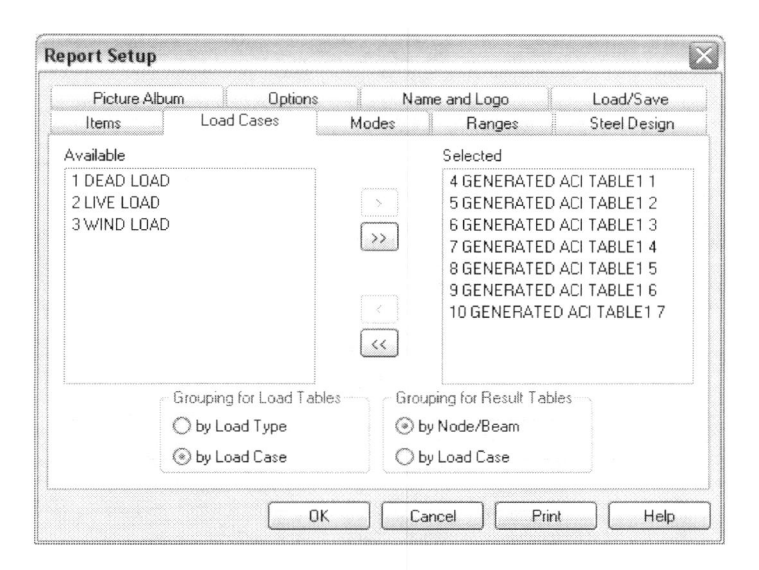

Ranges tab
- In this tab, user specifies if the report should contain results about all Nodes, Beams, and Plates, or some of them.

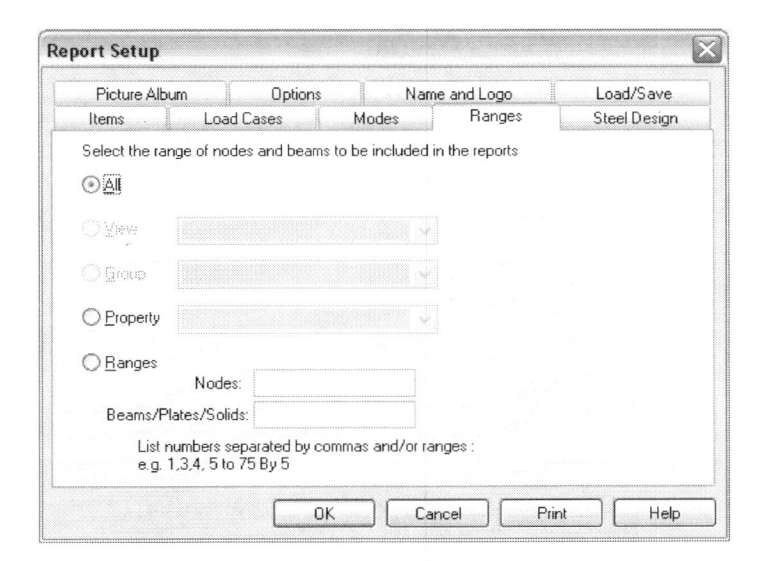

Picture Album tab
- In this tab, user can browse the pictures already capured.
- Specify whether to delete any of them.
- You can leave each picture with it's defualt size or:
  - Specify if you want each picture to take a Full Page.
  - Instead specify the Height and Width of each picture.

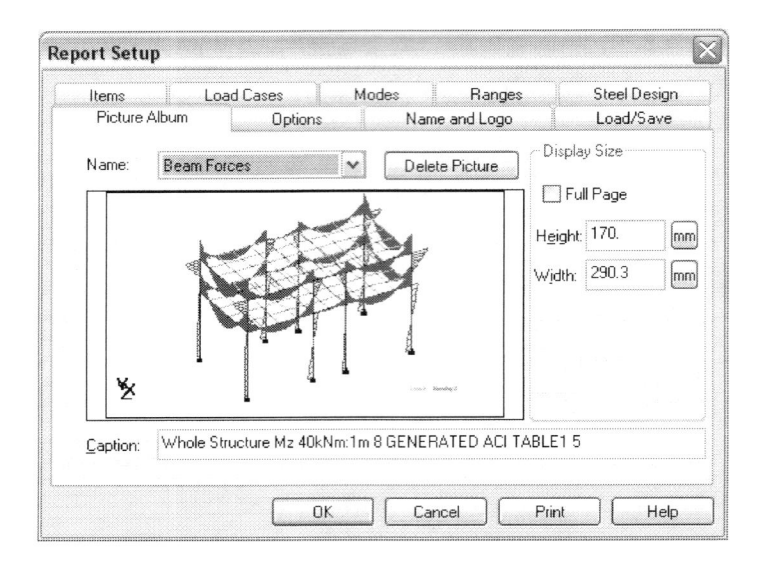

Options tab
- In this tab, you will specify the general options of the report, which will control its looks.

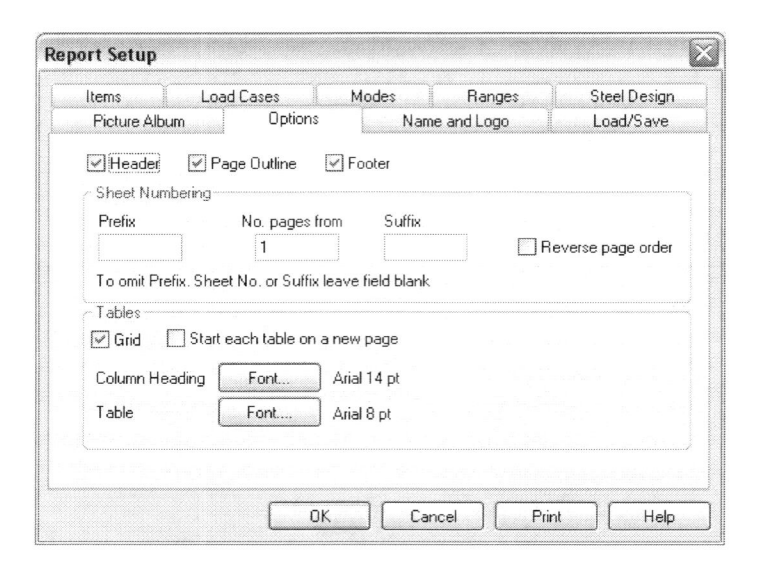

- Specify to show or hide:
  - Header
  - Footer
  - Page Outline (the frame appears around the page)
- In the **Sheet Numbering** part, specify the following:
  - To include a **Prefix** (for instance the word *Page*) or not
  - The staring page number of the report
  - To include a **Suffix**, or not
  - To Reverse page order, or not
- In the **Tables** part, specify:
  - Whether the table includes a grid, or without grid
  - Whether to start a new table in a new page, or not
  - Specify the Fonts for the column header, and contents of the tables

Name and Logo tab
- In this tab, you will specify the Name of the company you represent, and its logo, which will appear in all of the pages of the report.

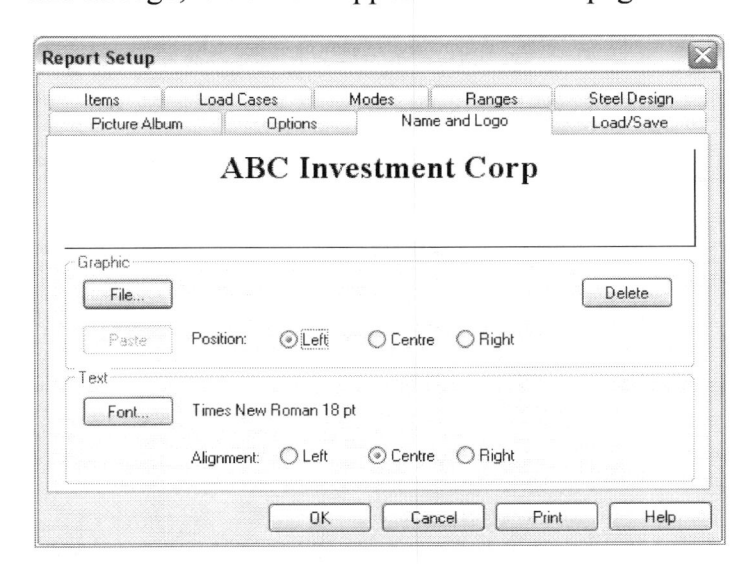

- In the white area, type the desired company name.
- In the **Graphic** part, click **File** button, and select the logo file of the company. Under the **Position** select the placement of the logo, whether Left, Center, or Right.
- In the **Text** part, select the **Font** to be used in the company name. Then select the palcement of it, as you did in the logo part.

Note
- If the position of the logo is centered STAAD.*Pro* will remove the company name. So it is better to keep the logo left always.

Load/Save tab
- In this tab you will be able to save the report contents.

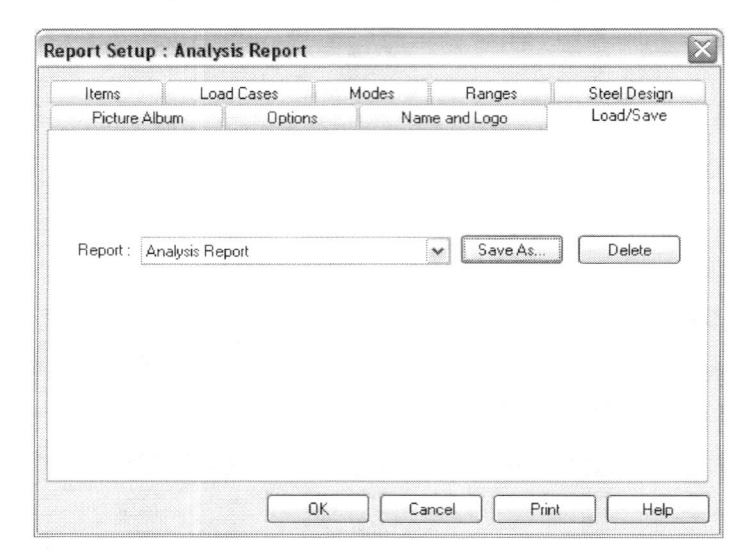

- Click the **Save As** button, the following dialog box will appear:

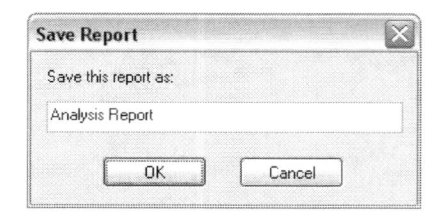

- Type in the name of the report (no need for any extension) and click **OK**.
- Next time you will find this file in the list, once you select it, it will be loaded.

Module 8: Post Processing

Print Preview Report
- By default the Print Preview mode in on.
- Other ways to view the report are:
    - From the Print toolbar, click the **Print Preview Report** icon.
    - From menus select **File/Print Preview Report**.
- On both ways you will get something similar to the following:

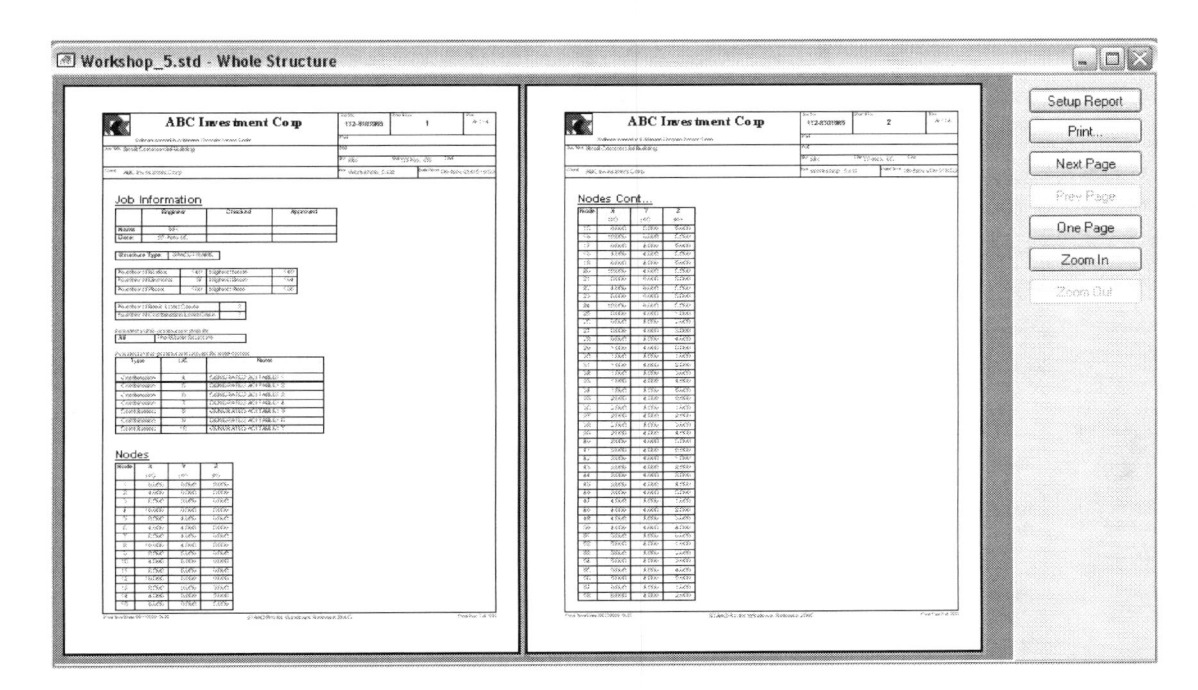

- Use **Zoom In**, and **Zoom Out**, to see the details of the report, or the whole picture.
- You can see **One Page**, or **Two Pages** simultaneously.
- Use **Next Page**, or **Prev Page** to navigate through the pages of the report.
- Use **Print** to print out the report as is.

Print Report
- In order to send the report to the printer, use of the following methods:
  - From the **Print** toolbar click the **Print** icon.
  - From menus, select **File/Print/Report**.
- In both ways you will see the following dialog box:

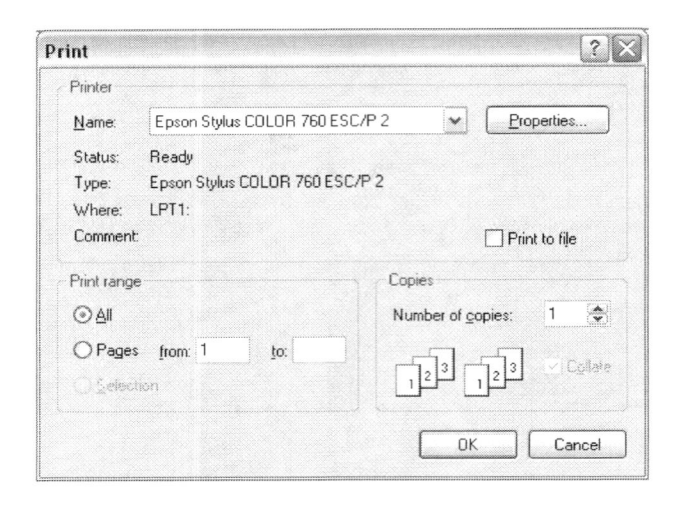

- Specify the following:
  - The Printer you want to send to.
  - The Print range, All, or certain page range.
  - The number of copies.

Export Report
- You can Export the saved report to:
  - Text File
  - MS Word File
- From menus select **File/Export Report**, and then choose **Text File** or **MS Word File**.

Note
- If you have a long report the process of exporting a report to MS Word file may take considerable time.

Module 8: Post Processing

Current View
- Another way of getting output from your input file is to deal with the current view displayed on the screen right now. You can:
  - Print Preview it
  - Print it
  - Export it
- Make sure that you are now displaying the current view (Displacement, Moment Diagram, Shear Diagram, etc…).
- From the **Print** toolbar click the **Print Preview Current View** icon. You will see the following picture:

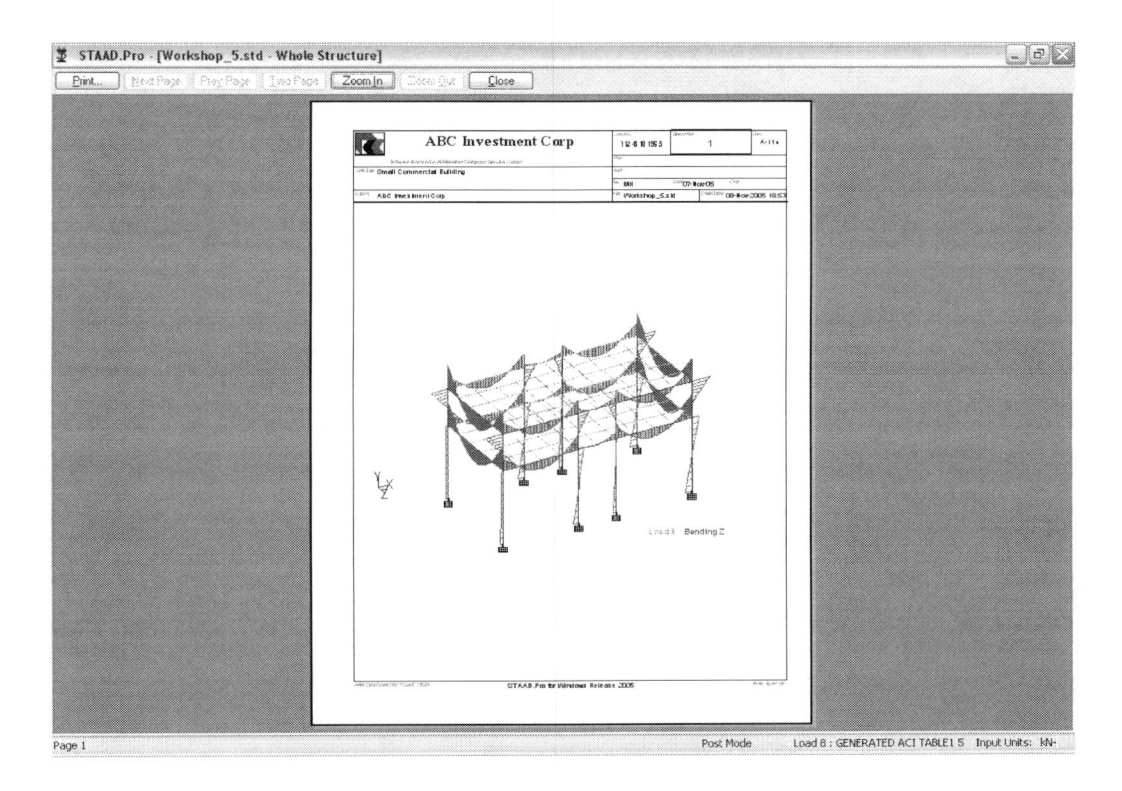

- From the **Print** toolbar click the **Print Current View** icon to print the current view. You will get the same dialog box of **Print Report**.

- From the **Print** toolbar click the **Export View** to save as the current view with one of the famous graphics file format such as jpg, tif, bmp, etc.

8-35

Other Ways: Double-Clicking a Beam

- To see the results of an individual Beam, simply double-click it.
- Make sure that you are using the Beam Cursor.
- Double-click on the desired Beam, you will see the following dialog box:

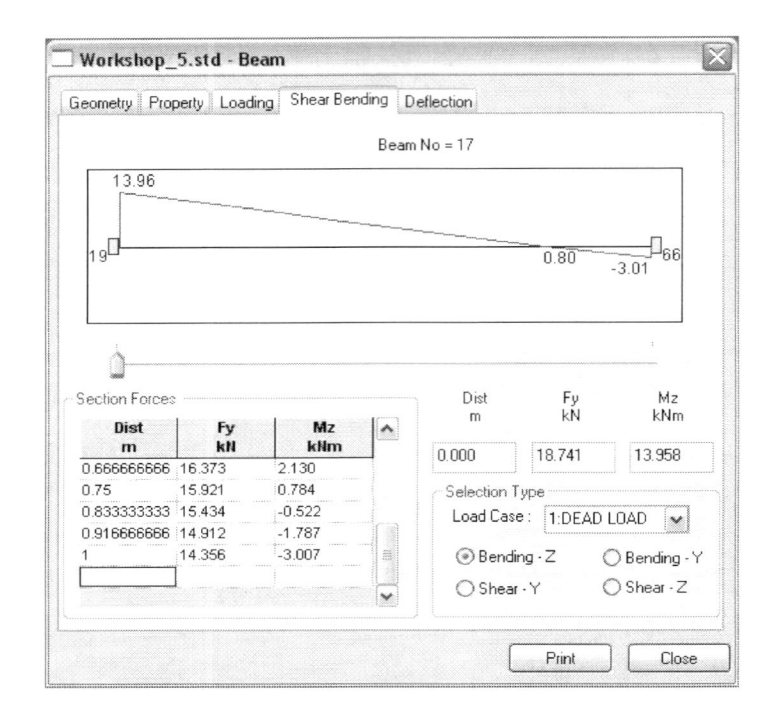

Shear Bending tab
- From the **Selection Type** part, decide on the following:

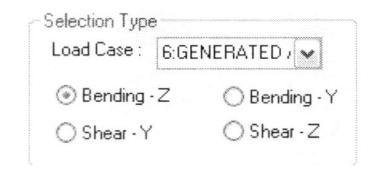

  - Desired **Load Case** to view the results for.
  - For the diagram type, you have four types to select from: Bending in Z direction; Shear in Y direction; Bending in Y direction; and finally Shear in Z direction. Most likely users will be interested in the first two items.

- Accordingly you will see the results in three different places:
  - Diagram view, which will show the Shear or Bending diagrams, with values at the ends, and distances at zero values

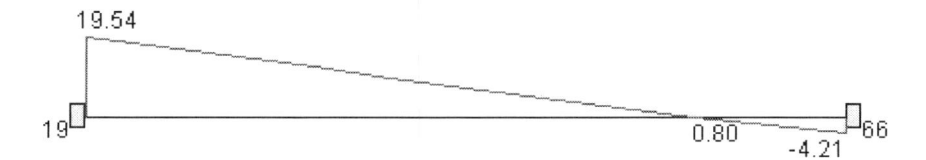

  - Table at the lower left portion of the dialog box, which will show always the values of Fy, and Mz at Intermediate and End points within the Beam selected

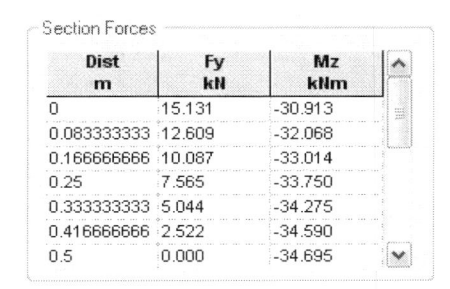

  - The slider, which will allow the user to move it to selected distances within the Beam, and hence see the corresponding values of Fy, and Mz.

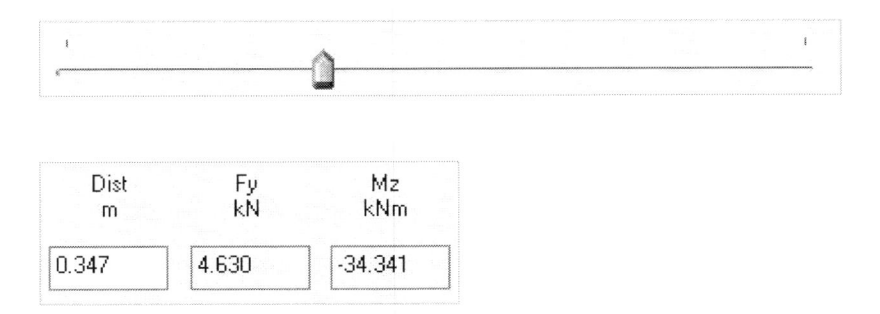

Deflection tab ■ If you select the **Deflection** tab, you will see the following dialog box:

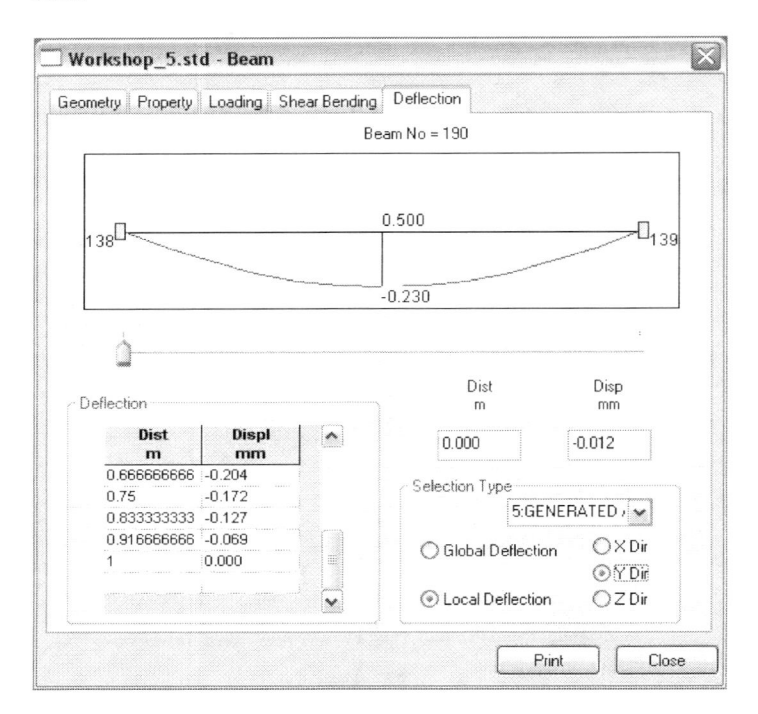

■ The same above procedure applies.

Note ■ User can print all screens, simply click **Print** button.
■ Also use menus, **Tools/Query/Member**, after you select a Beam.

Other Ways: Double-Clicking a Plate

- To see the results of an individual Plate, simply double-click it.
- Make sure you are using the Plate Cursor.

Center Stresses tab
- Click the **Center Stresses** to see the following dialog box:

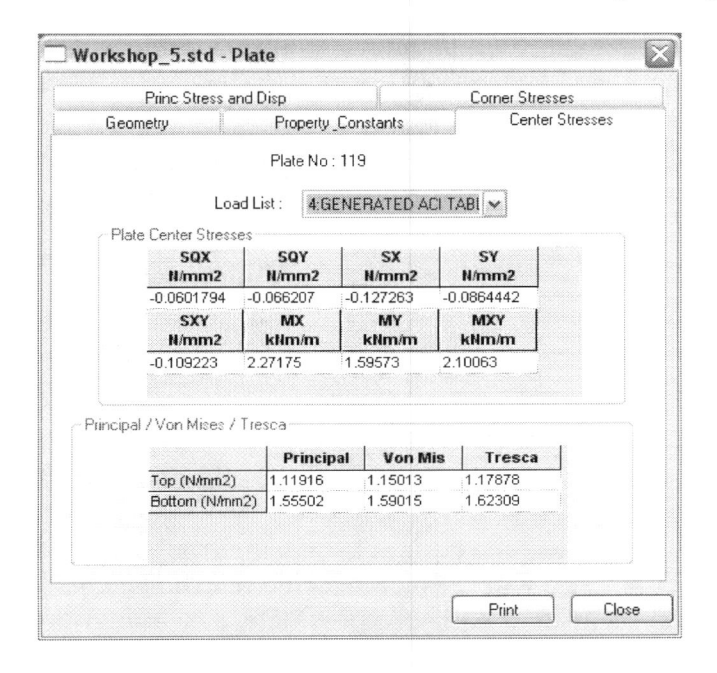

Workshop_5.std - Plate

Tabs: Princ Stress and Disp | Corner Stresses | Geometry | Property_Constants | **Center Stresses**

Plate No : 119

Load List : 4:GENERATED ACI TABL

Plate Center Stresses:

SQX N/mm2	SQY N/mm2	SX N/mm2	SY N/mm2
-0.0601794	-0.066207	-0.127263	-0.0864442

SXY N/mm2	MX kNm/m	MY kNm/m	MXY kNm/m
-0.109223	2.27175	1.59573	2.10063

Principal / Von Mises / Tresca:

	Principal	Von Mis	Tresca
Top (N/mm2)	1.11916	1.15013	1.17878
Bottom (N/mm2)	1.55502	1.59015	1.62309

- Select the desired Load Case from the **Load List**.
- In this tab, you will see the Plate Center Stresses, SQX, SQY, SX, SY, SXY, MX, MY, and MXY.
- Also, you will see the Principal, Von Mis, and Tresca stresses at the top and bottom of the Plate.

Princ Stress and Disp tab

- Click the **Princ** (Princ is an abbreviation for Principal) **Stress and Disp** to see the following dialog box:

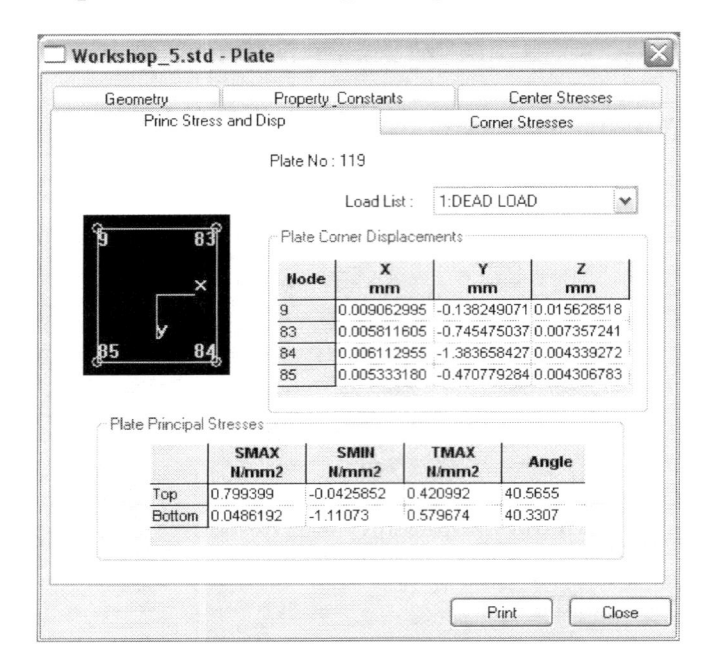

- Select the desired Load Case from the **Load List**.
- You will see picture at the left showing the number of the Nodes at the four corners of the Plate. Accordingly you can read from the **Plate Corner Displacements** table Displacement of each Node in X, Y, and Z Axis.
- Also, you can see the **Plate Principal Stresses** which they are; SMAX, SMIN, and TMAX at the top and bottom of the Plate.

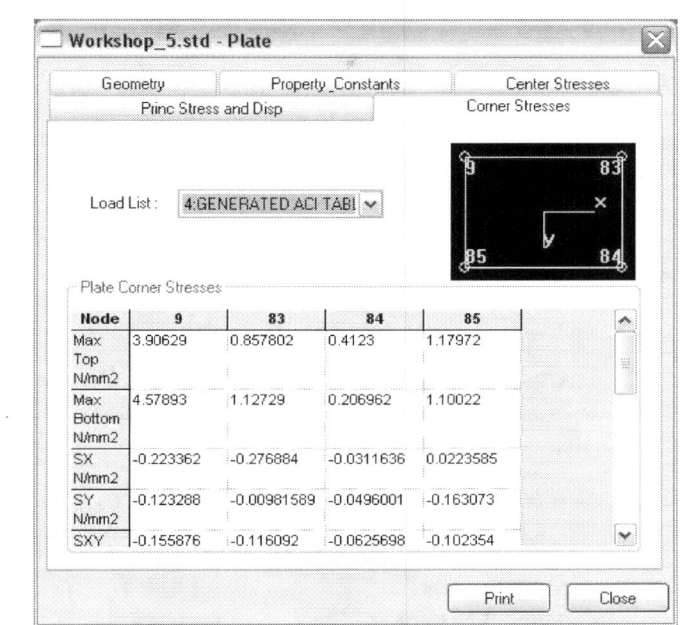

Corner Stresses tab ■ Click the **Corner Stresses** to see the following dialog box:

■ Select the desired Load Case from the **Load List**.
■ You will see picture at the left showing the number of the Nodes at the four corners of the Plate. Accordingly you can read from the **Plate Corner Stresses** table all the types of stresses at these plate corners.

Note ■ User can print all screens, simply click **Print** button.
■ Also use menus, **Tools/Query/Plate**, after you select a Plate.

Seeing the Results

Workshop 6-A & 6-B

1. Open Small_Building file.
2. Go to the **Post Processing Mode**. Choose only the Combinations.
3. Browse through the different Pages, and sub-pages of the Post Processing; see the tables and the graphs. Change the Scale of the different results as needed.
4. While you are doing so, **Take Picture**s of some of the results.
5. Create a report, which includes Input data, Output data, and some of the taken pictures.
6. Re-arrange the contents according to your desire.
7. Specify a Name and a Logo to appear in the report.
8. Print Preview the Report.
9. If you like it, print two page of it.
10. Close without saving.

Module Review

1. The new method of output in STAAD.*Pro* is better than old STAAD-III methods, but these old methods still available in STAAD.*Pro*:

    a. True

    b. False

2. The Load cases to appear in the tables of the Post Processing Mode are:

    a. Dictated by STAAD.*Pro*

    b. Can be customized by the user once selecting the Post Processing Mode

    c. Always the Combinations

    d. None of the above

3. Press _____ to see the Animation in full screen window

4. Reports can be exported to XLS files

    a. True

    b. False

5. In which of the following methods I can view the Shear/Bending diagrams with the distances along the Beam with ZERO Shear/Bending:

    a. From Node Displacement tab

    b. From Beam Forces tab

    c. From Double Clicking the Beam

    d. None of the above

6. Use from the menus _____ to show results values on the diagrams

Module Review Answers

1. a
2. b
3. F12
4. b
5. c
6. Results/View Value

Module 9:

Concrete Design

This module contains:

- Modes of the new Concrete Design Engine
- Steps to accomplish the Design Process
- Steps to view and verify the Design Results for Beams
- Steps to view and verify the Design Results for Columns
- Producing the Design Report

Introduction

- Concrete Design in STAAD.*Pro* has been improved significantly.
- Since STAAD.*Pro* 2001, a new separate-but-integrated RC Designer module has been added, which reads the Analysis file, and perform the design for concrete columns, and beams.
- Hence, analysis is the first step of the Concrete Design.
- From menus select Mode/Interactive Designs/Concrete Design.
- STAAD.*Pro* will take the user to a new program called **STAAD.*Pro* RC Designer**, the screen will change to:

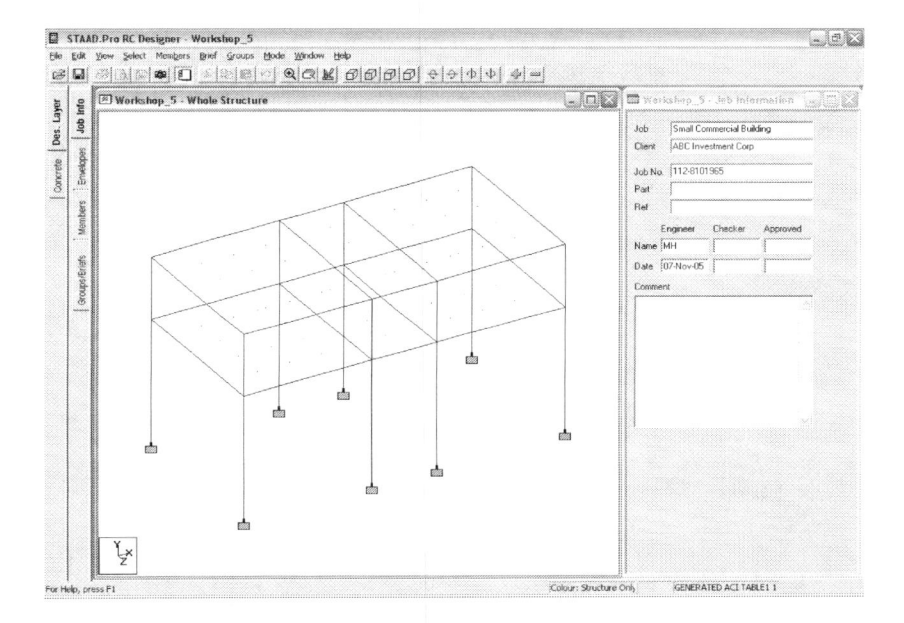

- This screen is almost identical in its layout to the STAAD.*Pro* screen.

Modes of Concrete Design

- Concrete Design got two modes:
  - Design Layer Mode
  - Deign Mode
- In Design Layer Mode, you specify the inputs of the concrete design process.
- Then user issues the command of design.
- In Design Mode, you will see the results, and generate reports.
- Just like in STAAD.*Pro*, for each mode, there will be different Page Control, and sub-pages.
- Also, for each Page Control, and sub-page, there will be a different Data Area.
- You will be in the Design Layer Mode by default.
- Here are the steps you will undergo to accomplish the designing process.

Step 1: Job Info

- By default the Job Info of your Analysis file will be transferred to your Concrete Design file.
- You can change the Job Info as you wish in the Concrete Design.

Module 9: Concrete Design

Step 2: Creating Envelopes

- Click on the **Envelopes** Page Control.
- In this Page Control you will specify the Primary and/or the Combination Loads, which will be involved in the design process.
- At the Data Area you will see the following:

- Click on the **New Env.** button to create a new Envelope, the following dialog box will appear:

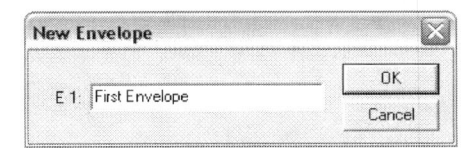

- Type in the name of the Envelope and click **OK**.
- The following dialog box will appear:

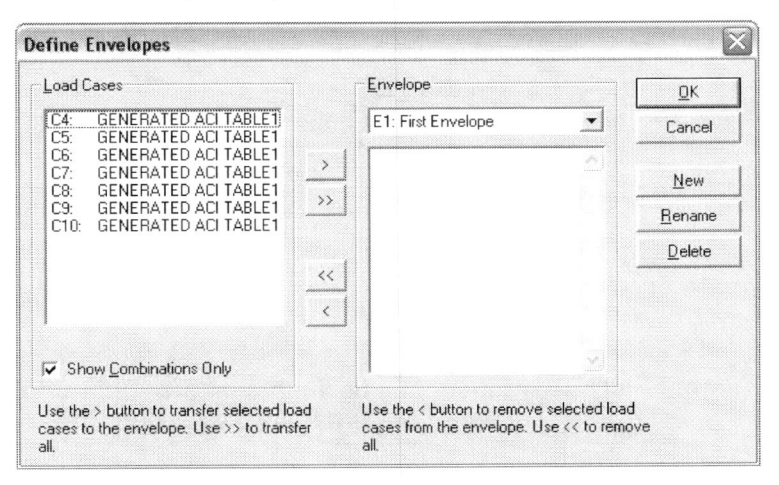

- By default only Combinations are shown, to show even the Primary Loads, simply click OFF the check box **Show Combinations Only**.
- You will see the dialog box changes to the following:

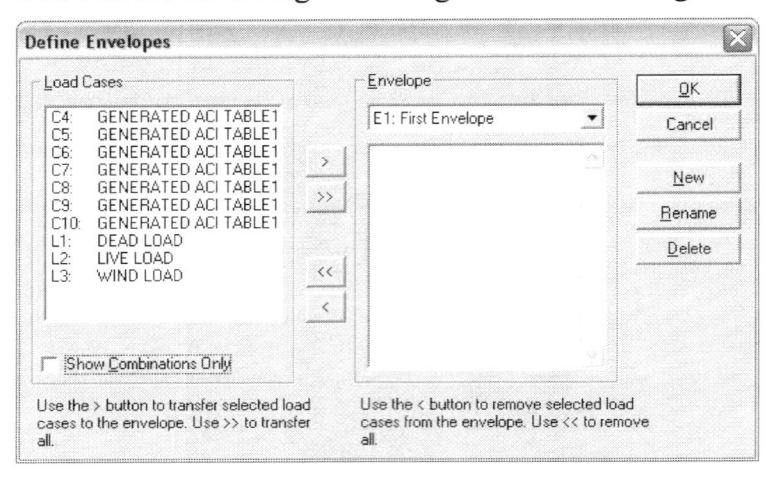

- Now select the desired Primary and/or Combination loads and click one arrow to the right button, or click the two arrows to the right button to select all the loads. Click **OK**. 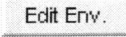
- To edit the Envelope, simply select the envelope from the table, and click **Edit Env.** button. The same dialog box will reappear, so you can make the desired changes.

Step 3: Creating Members

- Click on the **Members** Page Control.
- This page is specifically important to create continuous beams for designing purposes.
- STAAD.*Pro* treats any member between two Nodes as a separate Beam, hence adjacent Beams (which are separated by columns) will be considered, as *not related* Beams, and this is wrong assumption in concrete design.
- **Members** Page Control came to solve this problem.
- To create a Member, you have two ways:
  - Form Member
  - Auto Form Member

Form Member
- This function is usefeul to create continuous beams.
- Select the desired Beams you want to form as a single Member.

Module 9: Concrete Design

- From toolbars click **Form Member** button.
- Automatically the selected Beams will form a single Member named as M1 (if this is the first time you define Members in this file).

- Accordingly the Data Area will show the following information:

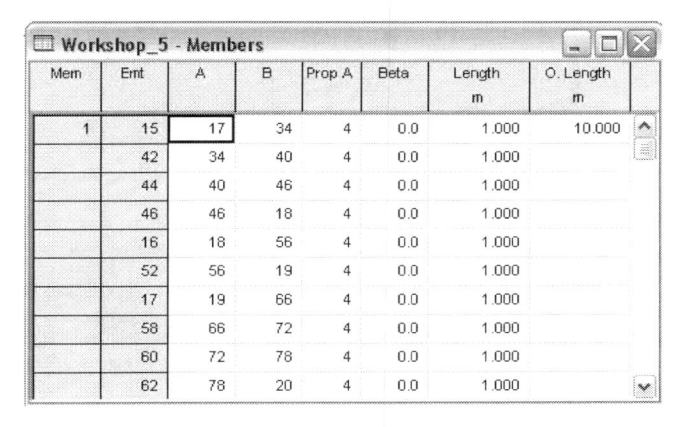

Mem	Emt	A	B	Prop A	Beta	Length m	O. Length m
1	15	17	34	4	0.0	1.000	10.000
	42	34	40	4	0.0	1.000	
	44	40	46	4	0.0	1.000	
	46	46	18	4	0.0	1.000	
	16	18	56	4	0.0	1.000	
	52	56	19	4	0.0	1.000	
	17	19	66	4	0.0	1.000	
	58	66	72	4	0.0	1.000	
	60	72	78	4	0.0	1.000	
	62	78	20	4	0.0	1.000	

- In this table you will see the following data:
  - The Member number
  - The number of the Beams (here they call it Emt which is an abbreviation of Element) as defined by STAAD.*Pro* input file
  - The number of the Nodes of the two ends of each Beam
  - The Properties (Cross-Section) of the beginning and end Nodes of the Beams
  - Beta angle for each Beam (it should be consistent)
  - The length of each individual Beam.
  - The Overall length of all adjacent Beams.

Auto Form Member
- This function is important to create Column members.
- If you select multiple columns each will be a separate Member.

- From toolbars click **Auto Form Member** button.
- Automatically the selected Beams will form single Members and will be numbered as sequesnce M1, M2, M3, etc. or it will follow the last numbered Member.

- The Data Area will look something like the following:

Mem	Emt	A	B	Prop A	Beta	Length m	O. Length m
1	15	17	34	4	0.0	1.000	10.000
	42	34	40	4	0.0	1.000	
	44	40	46	4	0.0	1.000	
	46	46	18	4	0.0	1.000	
	16	18	56	4	0.0	1.000	
	52	56	19	4	0.0	1.000	
	17	19	66	4	0.0	1.000	
	58	66	72	4	0.0	1.000	
	60	72	78	4	0.0	1.000	
	62	78	20	4	0.0	1.000	
2	7	1	5	2	0.0	4.000	4.000
3	21	13	17	2	0.0	4.000	4.000

Step 4: Creating Briefs

- Click on the **Groups/Briefs** Page Control.
- We will discuss now only the **Briefs** part, and in the next step we will cover the **Groups** part.
- In the **Briefs** we will set the values of the design requirements, which belong to this specific design case. There are **Beam Brief**, and **Column Brief**.
- The Briefs we will discuss is for **ACI**.

New Brief
- From Data Area click the **New Brief** button at the lower part of the screen.

Beam Brief
- The following dialog box will appear:

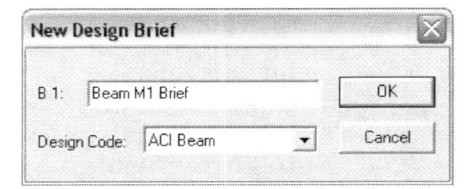

New Design Brief

B 1: Beam M1 Brief

Design Code: ACI Beam

9-8

- Type in the name of the Brief.
- From **Design Code** pop-up list select **ACI Beam**.
- Click **OK**.

■ In the Data Area you will see the following:

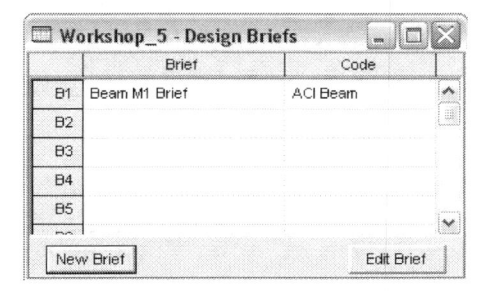

■ You can create as many Briefs as you wish.
■ You can create for each Beam a separate Brief.

Edit Brief

■ To change (edit) the values in any **Brief**, simply select this Brief from the list, and click **Edit Brief** button.
■ The following dialog box will appear:

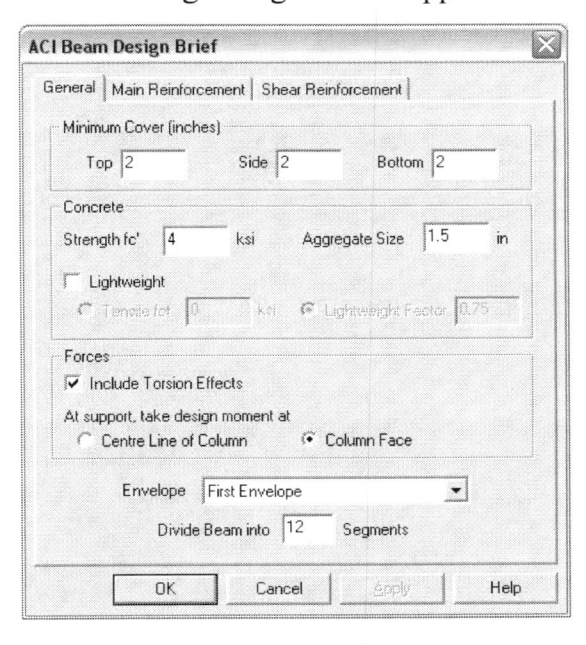

■ In the **ACI Beam Design Brief** there are three tabs, which they are:
- General
- Main Reinforcement
- Shear Reinforcement

General
- Specify the **Minimum Cover** from the 3 sides: Top, Side and Bottom (*Default value = 2 in*).
- Specify the design concrete strength f_c' (*Default = 4 ksi*)
- Specify the Maximum **Aggregate Size** to be used in the Concrete Mix (*Default value=1.5 in*)

- Also, you can specify if a **Lightweight** concrete or not, if yes, specify the criteria to be used; either:
  - Input the tensile strength as a value (using ksi)
  - Or you can input it as a percentage of the compressive strength

- Select to Include Torsional Effects or not?
- Specify whether to design for moments at **Center Line of Column**, or **Column Face** (*Default*).
- Select the desired **Envelope** from the provided pop-up list (this is the only place to link the previously made Envelope and the Brief).
- To calculate the max Shear and Bending values, specify the number of segments of the Beam (*Default value=12 segments*)

Main Reinforcement
- Click on the **Main Reinforcement** tab, you will see:

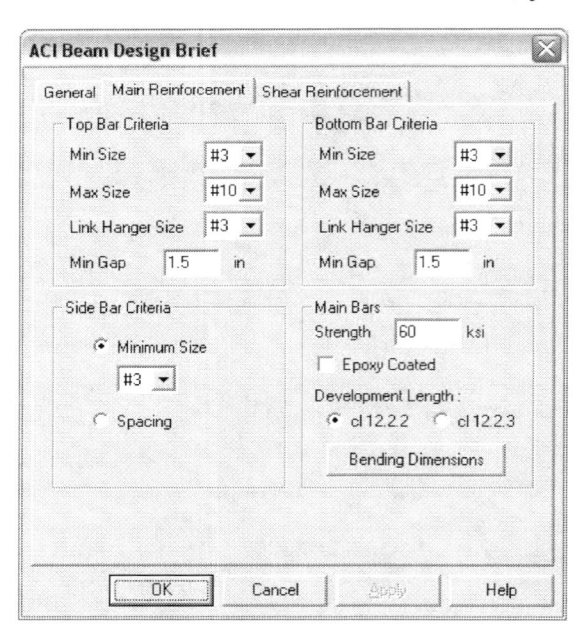

- In the **Top Bar Criteria** section, you will specify:
  - Min. Size of steel to be used (*Default=# 3*)
  - Max. Size of steel to be used (*Default=#10*)

- In case of there is no need for steel to resist bending but to be as supporting links, select the **Link Hanger Size** (*Default = #3*).
- In case of more than one layer of steel in the top, provide the **Min. Gap** (*Default = 1.5 in*).

■ The same applies to the **Bottom Bar Criteria** section.
■ In case of Deep Beams, the **Side Bar Criteria** section will control whether:
- You will specify the **Minimum** Bar **Size** of steel (*Default = # 3*).
- **Spacing**, which means will leave it for program to find the suitable bar size.

■ In the **Main Bar Type** section, you either:
- Input bar **Strength** (*Default = 60 ksi*), you can also set if **Epoxy Coated**, or not?
- Specify whether to calculate **Development Length** using clause 12.2.2, or clause 12.2.3. Also, click the **Bending Dimensions** button to specify **Minimum Bending Radius** for each Bar Size.

Shear Reinforcement

■ Click on the **Shear Reinforcement** tab, you will see:

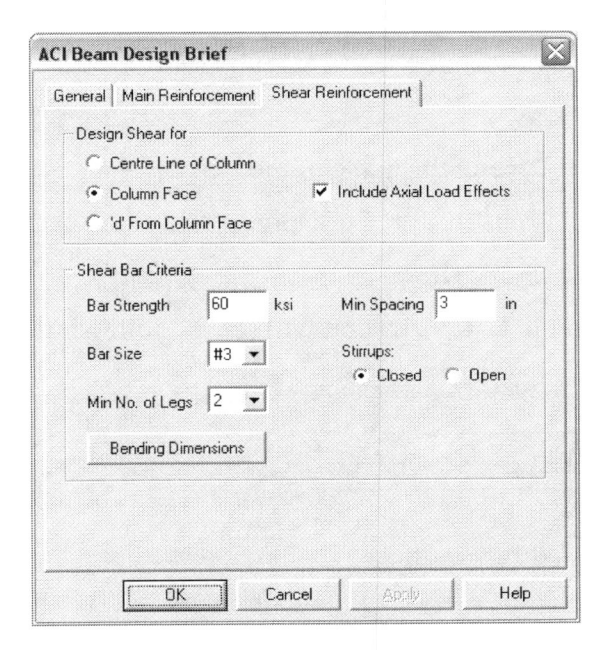

- In the Design Shear for section specify whether to design shear at Center Line of Support, Column Face, or 'd' From Column Face. In all cases, specify to Include Axial Load Effects, or not?
- In the **Shear Bar Criteria** section, control the following:
    - Steel **Bar Strength** (*Default = 60 ksi*)
    - Specify the bar **Size** will be used (*Default= #3*).
    - Specify the **Min. No. of Legs** for the stirrups (*Default=2*).
    - Specify the **Min. Spacing** between two stirrups (*Default=3 in*).
    - Specify the **Stirrups** type: **Closed** or **Open**.
    - Click the **Bending Dimensions** button to specify **Minimum Bending Radius** for each Bar Size.

Column Brief

- To create a Column Brief, from Data Area click the **New Brief** button.
- The following dialog box will appear:

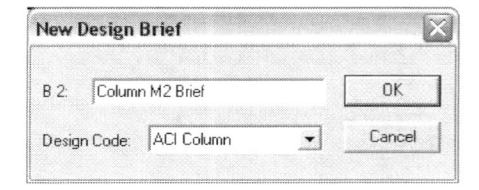

    - Type in the name of the Brief.
    - From **Design Code** pop-up list select **ACI Column**.
    - Click **OK**.
- In the Data Area you will see the following:

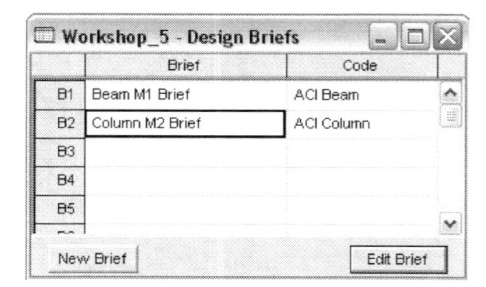

- You can create as many Briefs as you wish.
- You can create for each Column a separate Brief.

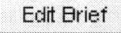

Edit Brief
- To change (edit) the values in any **Brief**, simply select this Brief from the list, and click **Edit Brief** button.
- The following dialog box:

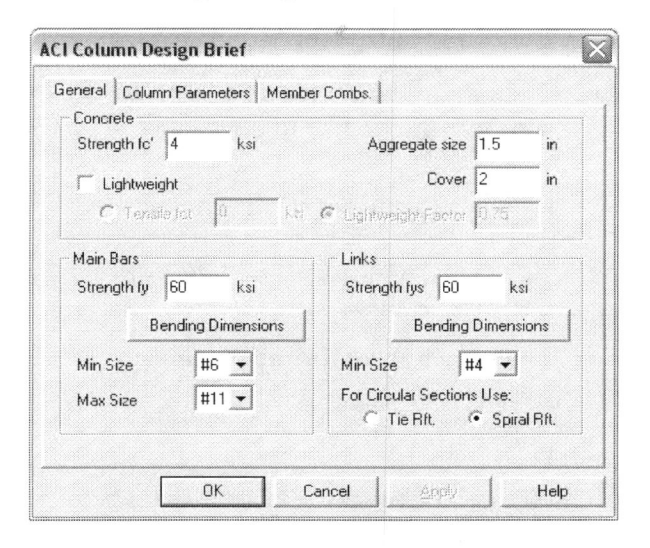

- As you can see in the **ACI Column Design Brief** there are three tabs, which they are:
  - General
  - Column Parameters
  - Member Combs

General
- In the Concrete section:
  - Specify the design concrete strength fc' (Default = 4 ksi)
  - Specify the Maximum Aggregate Size to be used in the Concrete Mix (Default value=1.5 in)
  - Specify the minimum Cover from Shear Links (Default = 2 in).
  - Also, you can specify if a Lightweight concrete used or not? if yes, specify the criteria to be used; either; input the tensile strength as a value (using ksi), or you can input it as a percentage of the compressive strength.

STAAD.Pro 2005 Tutorial

- In the **Main Bars** section, specify:
  - Input bar **Strength** (*Default = 60 ksi*).
  - Click the **Bending Dimensions** button to specify **Minimum Bending Radius** for each Bar Size.
  - Then input the **Minimum** bar size (*Default =# 6*), and **Maximum** bar size to be used (*Default =# 11*)
- In the **Links** section, specify:
  - Input bar **Strength** (*Default = 60 ksi*).
  - Click the **Bending Dimensions** button to specify **Minimum Bending Radius** for each Bar Size.
  - Then input the **Minimum** bar size (*Default = # 4*).
  - For Circular Section Use either; **Tie Rft** (Rft is Reinforcement), or **Spiral Rft**.

Column Parameters
- Click on the **Column Parameters** tab, and you will see the following dialog box:

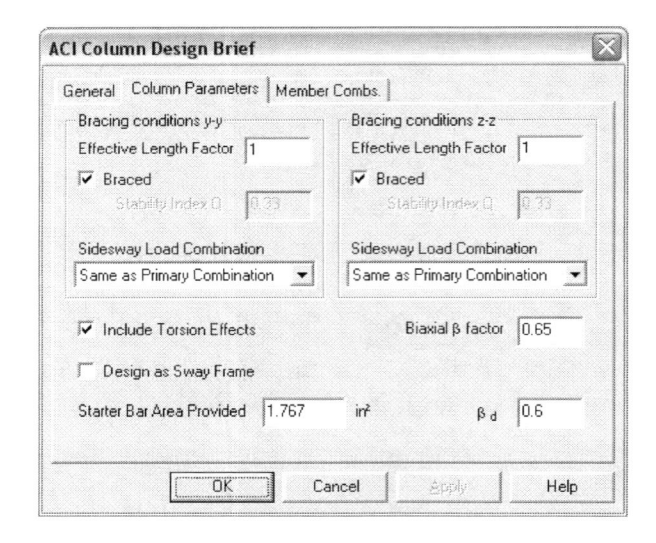

9-14

Module 9: Concrete Design

- In the **Bracing Conditions y-y** section specify
  - The Effective Length Factor (for the purpose of calculating Slenderness Ratio) (*Default = 1*)
  - Whether or not the Columns are braced in local y-y axis?
  - Select a specific Sidesway Load Combination, or set it to be Same as Primary Combination.
- The same applies for **Bracing Conditions z-z**.
- Specify to Include Torsional Effects, or not?
- Specify to **Design as Sway Frame**, or not?
- In the **Starter Bar Area Provided**, specify the Area provided for Starter bars, this is to calculate the final steel percentage.
- Specify Biaxial β factor, and $β_d$

Member Combs
- Click on **Member Combs** (Combs means Combinations) tab, and you will see the following dialog box:

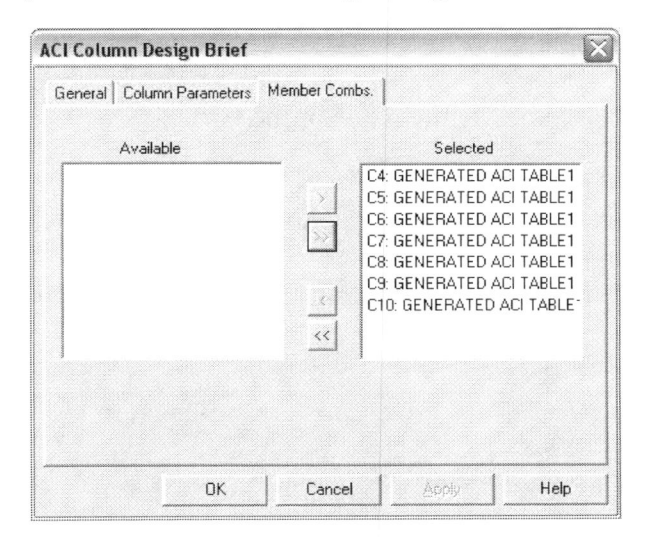

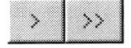

- Select the (already) made Combination, by using the one arrow or two arrows for all of the combination.

9-15

Step 5: Creating Groups

- You should be in the **Groups/Briefs** Page Control.
- Group is to combine the **Members** you created in Step 3, and the **Briefs** you created in Step 4.
- Follow the following steps:
  - As a first step select the desired Member using the current cursor. 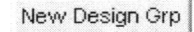
  - Now click **New Design Grp** (Grp means Group) button.
  - The following dialog box will appear: 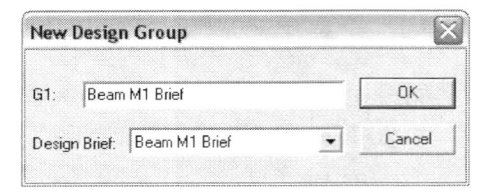
  - Specify the name of the Group.
  - Select the Design Brief. Click **OK**.
  - In the Data Area you will see the following:

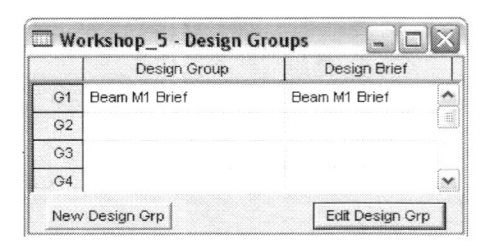

- Create other Groups as needed.
- In the **Page/Mode Automatic** toolbar you will see for each Group, there will be a member attached to it:

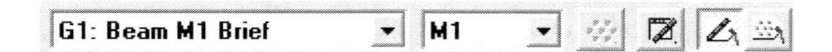

Module 9: Concrete Design

Step 6: Design Mode

- To activate all the inputs you made in the previous 5 steps, you have to go to the Design Mode.
- From menus select **Mode/Design**.
- The screen will change to the following:

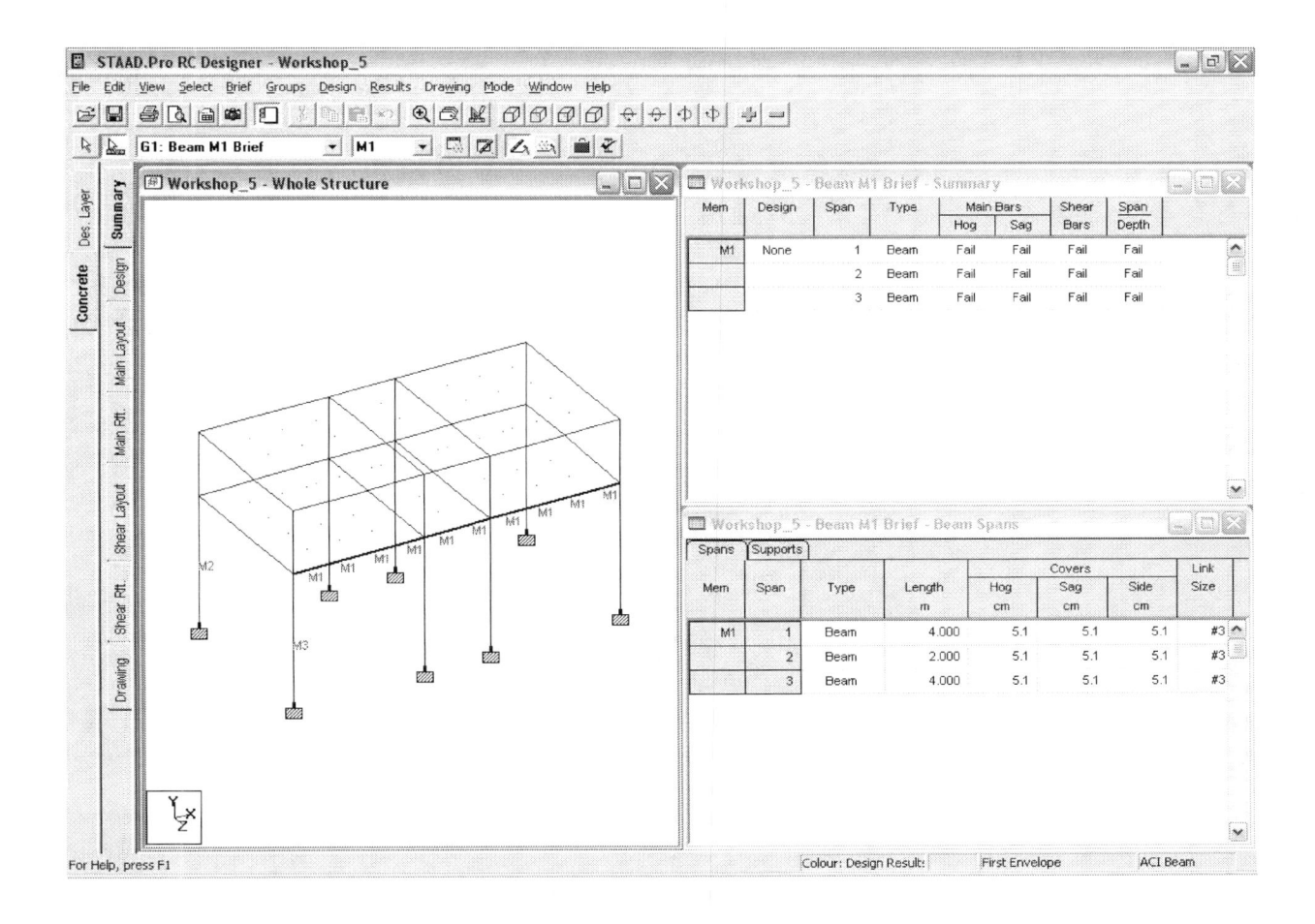

- At the beginning you will see that all results are **Fail**, don't worry this will change in a moment.
- By default last selected Group and Member in **Design Layer** Mode will be selected. You can select another Group and Members if you wish.

9-17

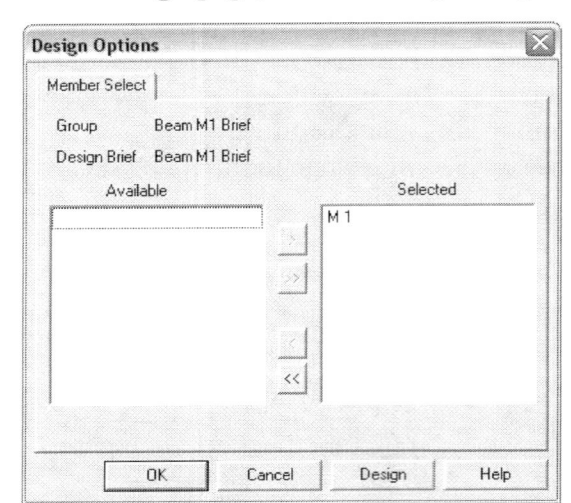

- Click **Design** page, the following dialog box will appear:

- You will find that the Members associated with the current Group are already selected.
- Click **Design** button.
- If there is any problem in the design, the below will appear:

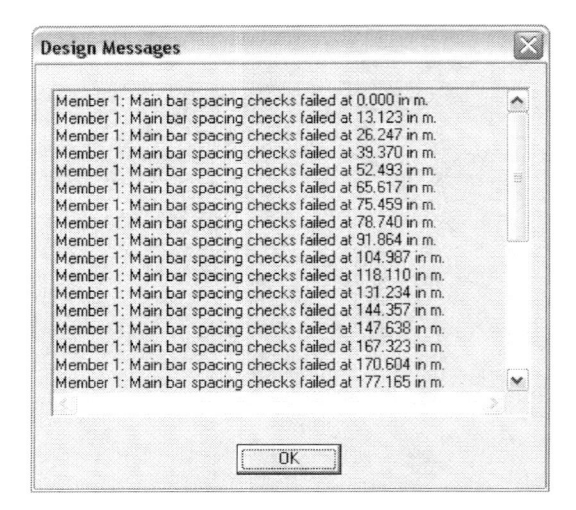

- In the case shown above you can see that STAAD.*Pro* is telling us that Main bar spacing checks fails for the whole length of the Beam.

Module 9: Concrete Design

- For Beam Design, by default you are in the **Summary** tab.
- In the Data Area, you will see the following two parts:
  - Summary table; which will list the Members, the Spans, the Main Bars status (Fail, or OK), the Shear Bars status (Fail, or OK), and the Span Depth status (Fail, or OK).

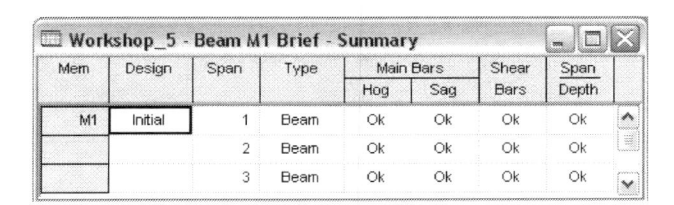

Mem	Design	Span	Type	Main Bars Hog	Main Bars Sag	Shear Bars	Span Depth
M1	Initial	1	Beam	Ok	Ok	Ok	Ok
		2	Beam	Ok	Ok	Ok	Ok
		3	Beam	Ok	Ok	Ok	Ok

  - Beam Span table which includes two tabs; the first one for the spans (Length, Covers, and Link size), and the second is for supports (the connection between Beam, and Column is considered Fixed Support).

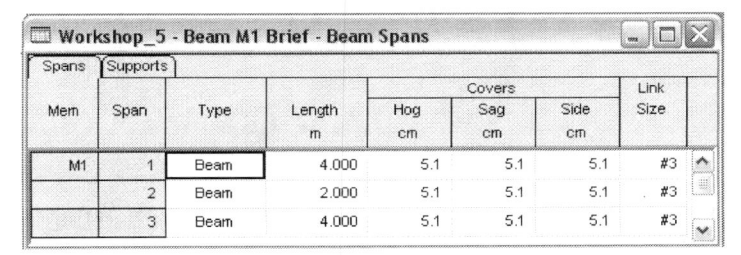

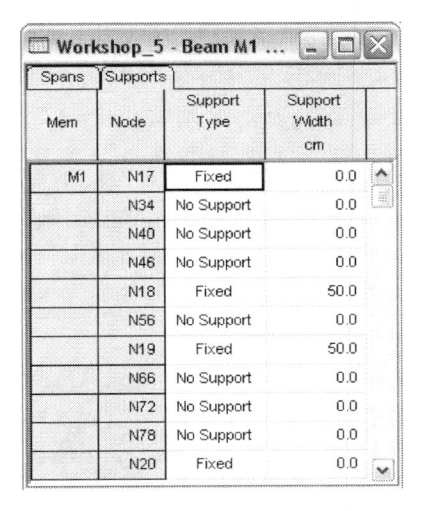

Spans tab:

Mem	Span	Type	Length m	Covers Hog cm	Covers Sag cm	Covers Side cm	Link Size
M1	1	Beam	4.000	5.1	5.1	5.1	#3
	2	Beam	2.000	5.1	5.1	5.1	#3
	3	Beam	4.000	5.1	5.1	5.1	#3

Supports tab:

Mem	Node	Support Type	Support Width cm
M1	N17	Fixed	0.0
	N34	No Support	0.0
	N40	No Support	0.0
	N46	No Support	0.0
	N18	Fixed	50.0
	N56	No Support	0.0
	N19	Fixed	50.0
	N66	No Support	0.0
	N72	No Support	0.0
	N78	No Support	0.0
	N20	Fixed	0.0

Step 7: Reading Results: Beam Main Layout

- Go to **Main Layout** page.
- In this page you will see – *for the selected Beam* – the following picture:

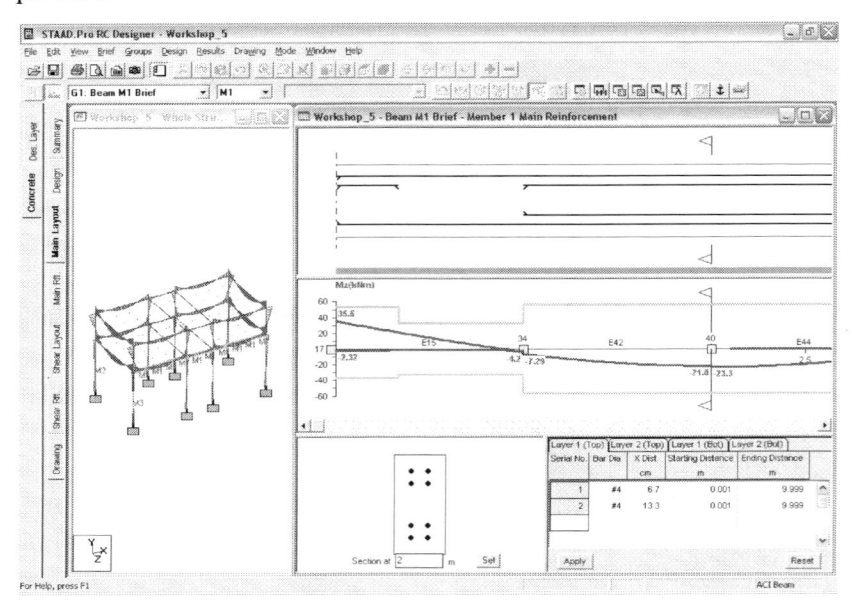

- This display is cut into four main synchronized views with single horizontal scroll bar:
    - The upper part shows the elevation view of the Beam with the reinforcement. You can also see the Plan view by right-click on the view and selecting **Plan View**.
    - The mid part shows the Moment Envelope.
    - The lower part to the left shows the cross section of the Reinforcement at specified distance from the Start of the Beam. Type in the section distance and click **Set** button, or, you can use the *Cross-Section indicator* in the upper part to move it to the desired location.
    - The lower part to the right shows a table outlining the layers of reinforcement top and bottom (in our example there are two layers at the top and two layers at the bottom), then the table lists where each bar starts and finishes along the member length.

Step 8: Reading Results: Beam Main Rft

- Go to **Main Rft** page (Rft means Reinforcement).
- In the Data Area you will see two tables, one of them, which is Beam Spans table, has already been discussed in the Summary tab.
- The other table is Main Reinforcement table, which includes two tabs; Hogging, and Sagging:
- Hogging is where bending moment is positive. Sagging is where bending moment is negative.
- In both tabs, you will see something similar the following table:

Distance m	Span	Moment kNm	As Req. mm²	As' Req. mm²	Bottom Layers Bars	Area mm²	Top Layers Bars	Area mm²
0.000	1(s)	35.493	325	0	2#4	253	2#4 2#4	506
0.333	1	21.504	205	0	2#4	253	2#4 2#4	506
0.667	1	8.464	222	0	2#4	253	2#4	253
1.000	1	0.000	0	0	2#4 2#4	506	2#4 2#4	506
1.333	1	0.000	0(110)	0	2#4 2#4	506	2#4 2#4	506

- Which includes:
  - Distance from the start of the Beam, and in which Span.
  - Moment value.
  - A_s Required.
  - A_s' Required.
  - Bottom Layers of Steel (Bars & Area)
  - Top Layers of Steel (Bars & Area)

Note
- In the table you will see the spans marked at the start of the span with the letter (s), and at the end of the span with letter (e).
- STAAD.*Pro* will search for the maximum positive moment at the beginning and at the end of each span, and then the maximum negative moment at intermediate points, accordingly will make the design based on these moments.

STAAD.*Pro* 2005 Tutorial

Step 9: Reading Results: Beam Shear Layout

- Go to **Shear Layout** page.
- In this page you will see – *for the selected Beam* – the following picture:

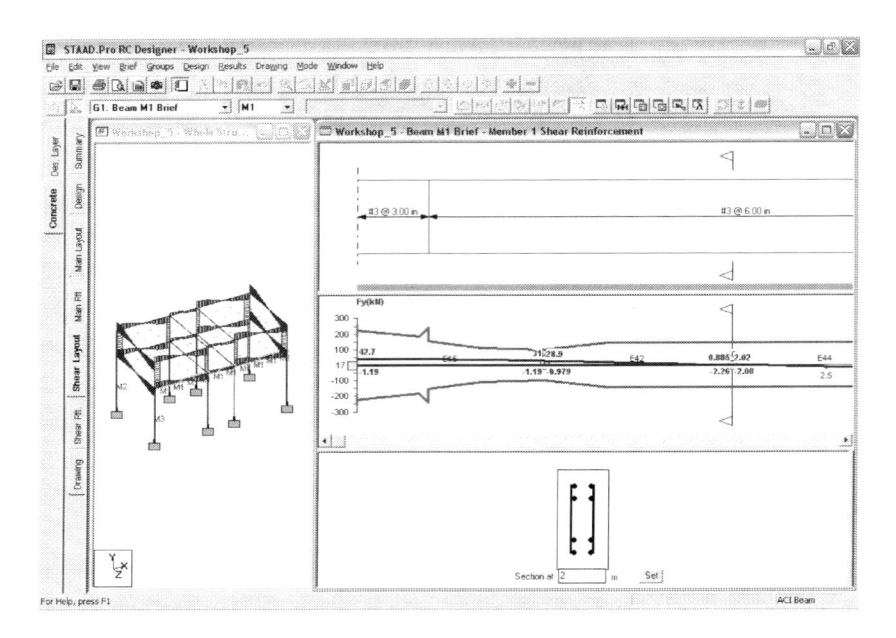

- This display is cut into three main synchronized views with single horizontal scroll bar:
  - The upper part shows the elevation view of the Beam with the shear reinforcement.
  - The mid part shows the Shear Envelope.
  - The lower part shows the cross section of the Shear Reinforcement at specified distance from the Start of the Beam. Type in the section distance and click **Set** button, or, you can use the *Cross-Section indicator* in the upper part to move it to the desired location.

9-22

Step 10: Reading Results: Beam Shear Rft

- Go to **Shear Rft** page (Rft means Reinforcement).
- In the Data Area you will see two tables. At the upper part you will see Shear Reinforcement table, which includes:

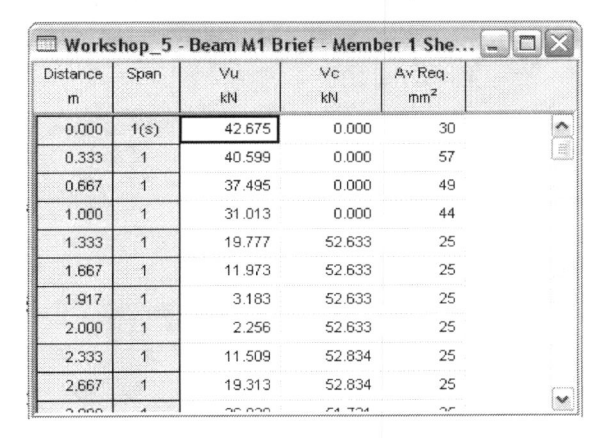

Distance m	Span	Vu kN	Vc kN	Av Req. mm²
0.000	1(s)	42.675	0.000	30
0.333	1	40.599	0.000	57
0.667	1	37.495	0.000	49
1.000	1	31.013	0.000	44
1.333	1	19.777	52.633	25
1.667	1	11.973	52.633	25
1.917	1	3.183	52.633	25
2.000	1	2.256	52.633	25
2.333	1	11.509	52.834	25
2.667	1	19.313	52.834	25

- Distance from the start of the Beam, and in which span.
- V_u value (in KN) at the location, V_c value (in KN) at the location, and finally the Area of Steel required (A_v in mm²).

- The other table is Shear Zones table:

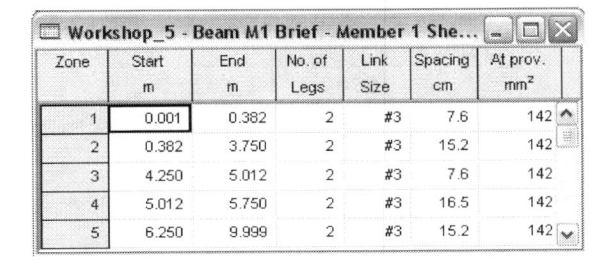

Zone	Start m	End m	No. of Legs	Link Size	Spacing cm	At prov. mm²
1	0.001	0.382	2	#3	7.6	142
2	0.382	3.750	2	#3	15.2	142
3	4.250	5.012	2	#3	7.6	142
4	5.012	5.750	2	#3	16.5	142
5	6.250	9.999	2	#3	15.2	142

- Which includes:
  - The Zone number, and where does it start along the member, and where does it finish.
  - Number of legs of the stirrup
  - The stirrup size
  - Spacing between stirrups (in cm).
  - The Area of Steel provided (in mm²)

Step 11: Reading Results: Beam Drawing

- You will find in the Data Area one table called Scheduled Bars:

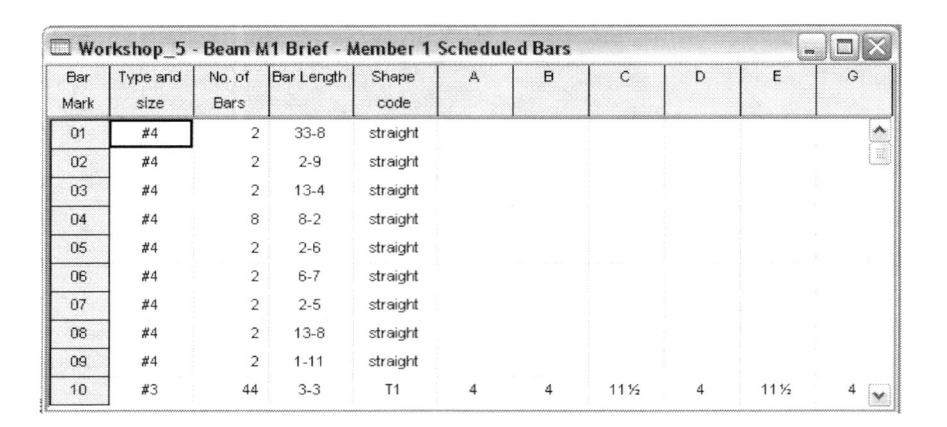

Bar Mark	Type and size	No. of Bars	Bar Length	Shape code	A	B	C	D	E	G
01	#4	2	33-8	straight						
02	#4	2	2-9	straight						
03	#4	2	13-4	straight						
04	#4	8	8-2	straight						
05	#4	2	2-6	straight						
06	#4	2	6-7	straight						
07	#4	2	2-5	straight						
08	#4	2	13-8	straight						
09	#4	2	1-11	straight						
10	#3	44	3-3	T1	4	4	11½	4	11½	4

- Which contains:
  - Bar Mark.
  - Type of the Bar, and its size.
  - How many bars to be used.
  - The bar length to be used.
  - Bar shape code according to ACI.
  - The A, B, C, D, etc. according to ACI.

Step 12: Reading Results: Column Main Layout

- Go to **Main Layout** page.
- In this page you will see – *for the selected Column* – the following picture:

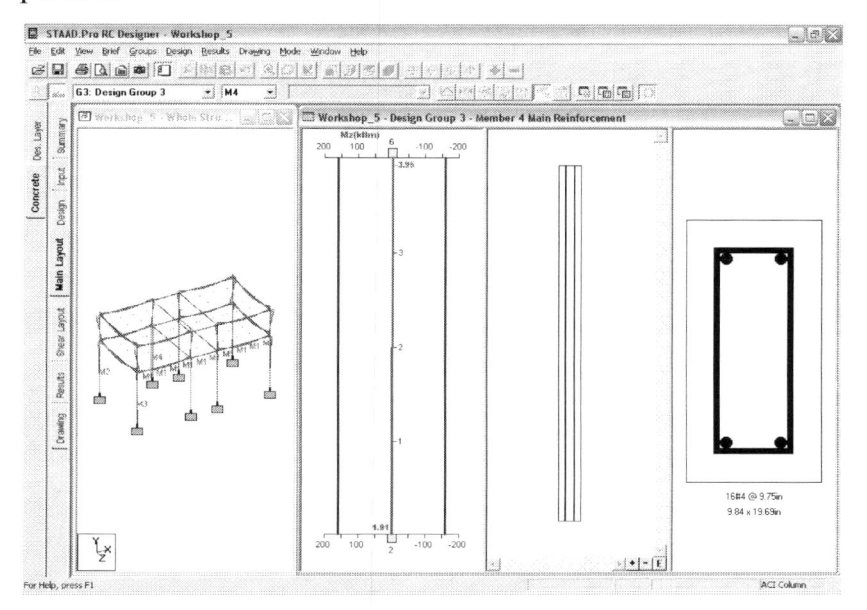

- In this view you will see:
    - Bending Moment diagram of the column
    - Elevation view of the column
    - Cross-sectional view of the column showing the main reinforcement steel bars.

Step 13: Reading Results: Column Shear Layout

- Go to **Shear Layout** page.
- In this page you will see – *for the selected Column* – the following:

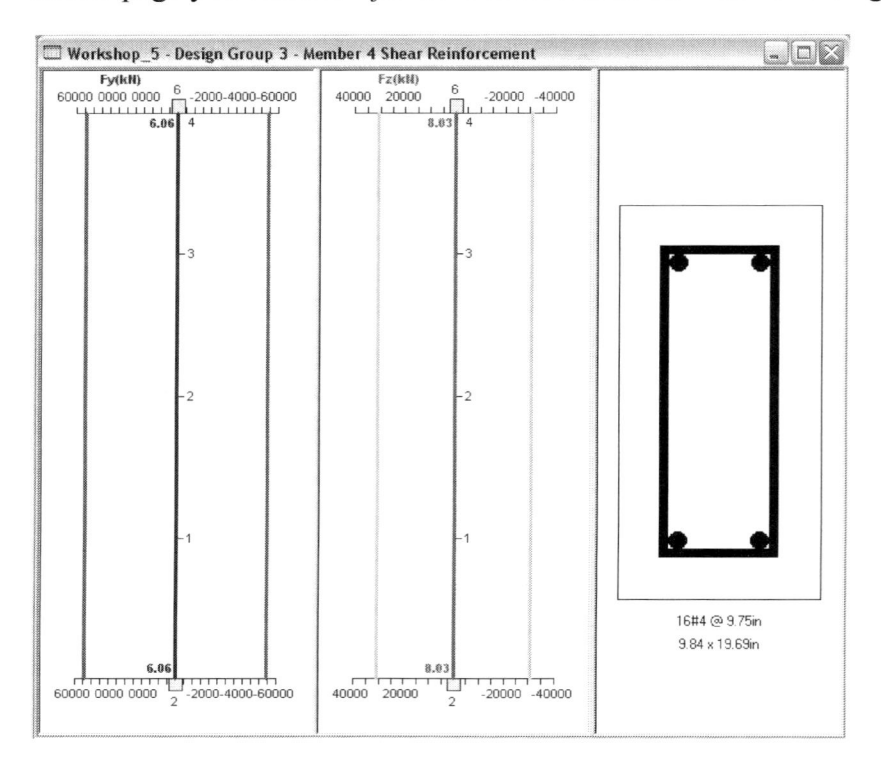

- In this view you will see:
  - Shear diagram of Fy
  - Shear diagram of Fz
  - Cross-sectional view of the column showing the shear reinforcement steel bars.

Step 14: Reading Results: Column Results

- Go to **Results** page.
- In the Data Area you will see two tables, at the upper part you will see Main Reinforcement table, which includes:

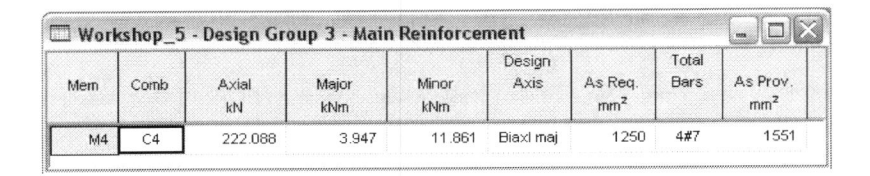

Mem	Comb	Axial kN	Major kNm	Minor kNm	Design Axis	As Req. mm²	Total Bars	As Prov. mm²
M4	C4	222.088	3.947	11.861	Biaxl maj	1250	4#7	1551

- Member number
- Combination number
- Axial Load value, Moment around Major axis, Moment around Minor axis, and which axis was designed for.
- Area of Steel required.
- Total number of bars suggested.
- Area of Steel provided.

■ At the lower part you will see Shear Reinforcement table, which includes:

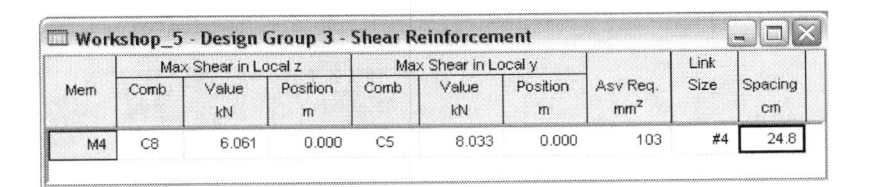

Mem	Max Shear in Local z Comb	Value kN	Position m	Max Shear in Local y Comb	Value kN	Position m	Asv Req. mm²	Link Size	Spacing cm
M4	C8	6.061	0.000	C5	8.033	0.000	103	#4	24.8

- Member number
- Max Shear in Local z & Local y, which includes Combination number, Shear value, and the position
- Area of Steel Required
- The Link size selected and Spacing between Links.

Step 15: Reading Results: Column Drawing

- You will find in the Data Area one table called Scheduled Bars:

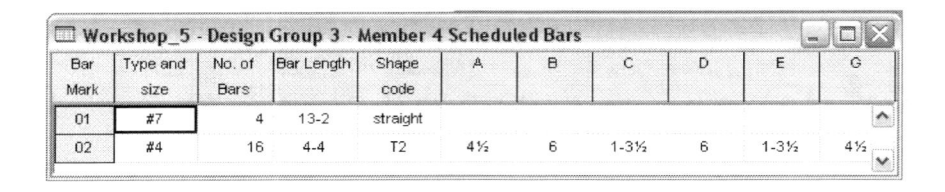

Bar Mark	Type and size	No. of Bars	Bar Length	Shape code	A	B	C	D	E	G
01	#7	4	13-2	straight						
02	#4	16	4-4	T2	4½	6	1-3½	6	1-3½	4½

- Which contains:
  - Bar Mark.
  - Type of the Bar, and its size.
  - How many bars to be used.
  - The bar length to be used.
  - Bar shape code according to ACI.
  - The distances A, B, C, D, etc. according to ACI.

Step 16: Reading Results: Generating Design Report

- The Design Report is very similar to Analysis Report, which we saw in Module 8, except this one shows the Design results and not the Analysis results.
- From the toolbars click **Report Setup** icon.
- The following dialog box will appear:

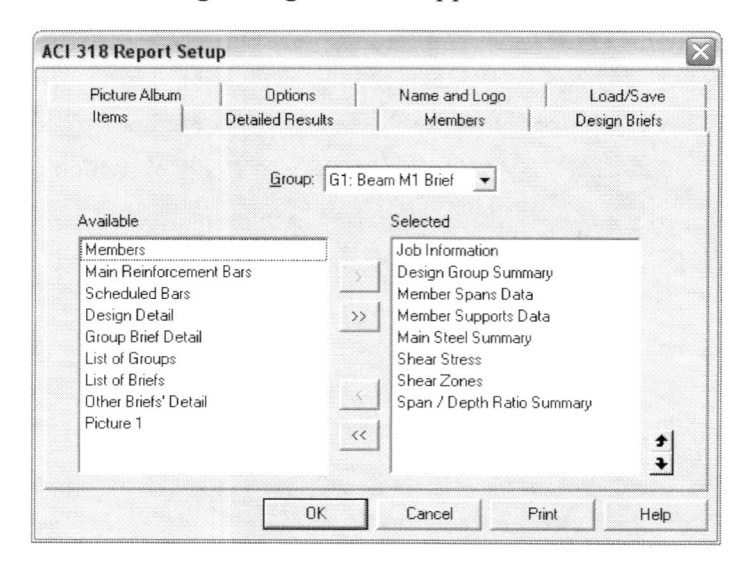

Items tab
- You will see that the current Group is already selected.
- In this tab you will select the data to be included inside the report.
- You can use the Arrow Up and the Arrow Down to change the location of each item in the report.

Detailed Results tab
- Click the **Detailed Results** tab, you will see the following dialog box:

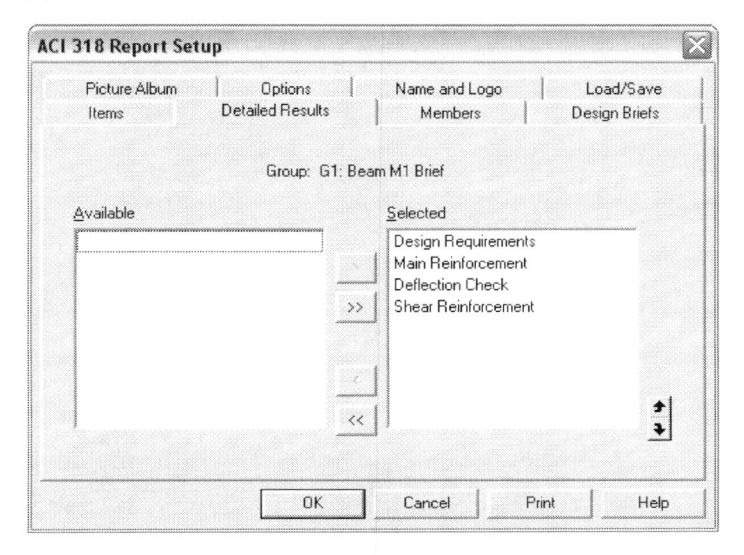

- By default all items are selected. They are: Design Requirements, Main Reinforcement, Deflection Check, and finally Shear Reinforcement.

Members tab
- Click the **Members** tab, you will see the following dialog box:

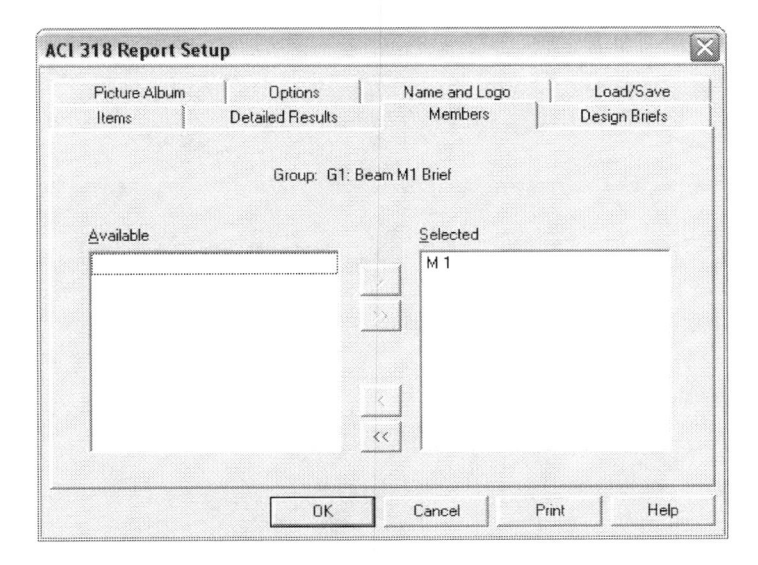

- Select which members design to be included in the report.

Design Briefs tab ■ Click the **Design Briefs** tab, you will see the following dialog box:

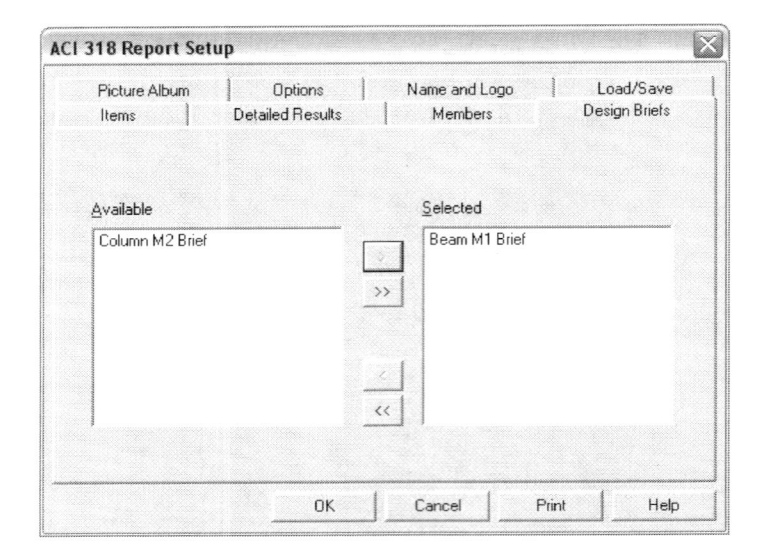

■ Select which Design Briefs will be included in the report.

Rest of tabs ■ Rest of tabs is identical to Module 8 Analysis Report part.

Module 9: Concrete Design

- To see the final report, from toolbar click **Print Preview** icon, or from menus select **File/Print Preview**.
- Both ways will lead to the following picture (this picture showing STAAD.*Pro* design calculation):

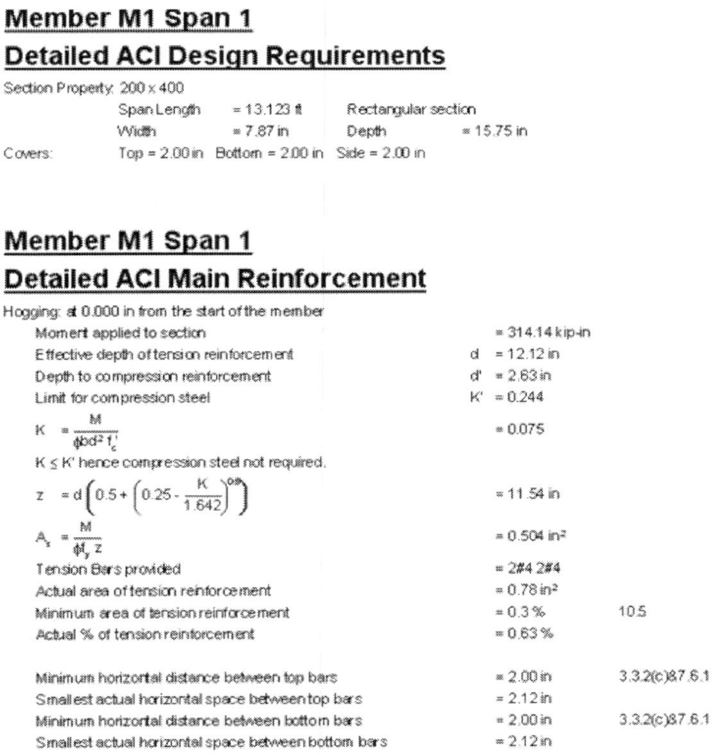

Member M1 Span 1
Detailed ACI Design Requirements

Section Property: 200 x 400
- Span Length = 13.123 ft   Rectangular section
- Width = 7.87 in   Depth = 15.75 in
- Covers: Top = 2.00 in   Bottom = 2.00 in   Side = 2.00 in

Member M1 Span 1
Detailed ACI Main Reinforcement

Hogging: at 0.000 in from the start of the member

Moment applied to section		= 314.14 kip-in
Effective depth of tension reinforcement	d	= 12.12 in
Depth to compression reinforcement	d'	= 2.63 in
Limit for compression steel	K'	= 0.244

$K = \dfrac{M}{\phi b d^2 f_c'}$ = 0.075

$K \leq K'$ hence compression steel not required.

$z = d\left(0.5 + \left(0.25 - \dfrac{K}{1.642}\right)^{0.5}\right)$ = 11.54 in

$A_s = \dfrac{M}{\phi f_y z}$ = 0.504 in²

- Tension Bars provided = 2#4 2#4
- Actual area of tension reinforcement = 0.78 in²
- Minimum area of tension reinforcement = 0.3 %   10.5
- Actual % of tension reinforcement = 0.63 %

- Minimum horizontal distance between top bars = 2.00 in   3.3.2(c)&7.6.1
- Smallest actual horizontal space between top bars = 2.12 in
- Minimum horizontal distance between bottom bars = 2.00 in   3.3.2(c)&7.6.1
- Smallest actual horizontal space between bottom bars = 2.12 in

- To send the report to the printer, from the toolbar click the **Print** icon, or from menus, select **File/Print**.
- In both ways you will see the following dialog box:

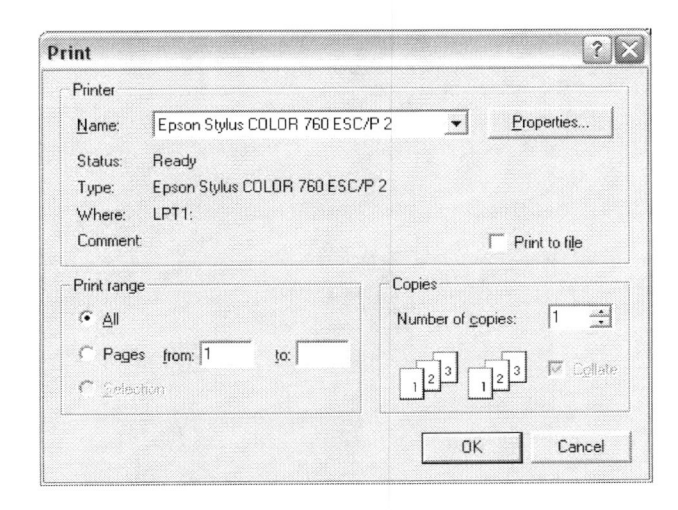

Concrete Design

Workshop 7-A

1. Open Small_Building file.
2. Select **Mode/Interactive Designs/Concrete Design**.
3. Define new Envelope call it **First Envelope** to contain all the loads (Primary and Combination).
4. From Members, define one beam, and one column as shown in the illustration below:

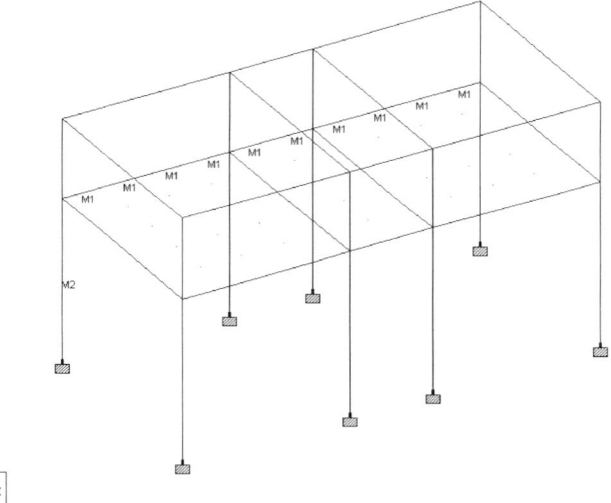

5. Create a new ACI Beam Brief, call it **Beam M1 Brief** and change the following values:

    a. **At Support, take design moments at** = Center Line of Column

    b. **Envelope** = First Envelope

    c. **Divide Beam into** = 4

    d. **Top Bar Criteria**: Min Size = #3, Max Size = #5

    e. **Bottom Bar Criteria**: Min Size = #3, Max Size = #5

    f. **Design Shear for**: 'd' From Column Face

    g. **Use Enhanced Shear Effects** = On

6. Create a new ACI Column Brief, call it **Column M2 Brief** and change the following values:

    a. **Main Bars**: Min Size = #8, Max Size = #14

    b. **Links**, For Circular Section Use = Tie Rft

    c. **Bracing conditions y-y**, Braced = ON

    d. **Bracing conditions z-z**, Braced = ON

    e. **Design as Sway Frame** = ON

    f. **Member Combs**, select all the available Combinations.

7. Select the Beam member.
8. From Design Groups, click **New Design Grp**, and create a new group and call it **First Group**, which contains the selected Beam member, and **Beam M1 Brief**.
9. Select the Column member.
10. From Design Groups, click **New Design Grp**, and create a new group and call it **Second Group**, which contains the selected Column member, and **Column M2 Brief**.
11. From menus select **Mode/Design**.
12. Make sure that the upper part of the screen is showing **First Group** and **M1**.
13. From Page Control, click **Design**, a dialog box will appear, select **M1**, and click **one arrow to the right**, then click **Design**.
14. Look at the results.
15. From the upper part of the screen switch to **Second Group** and **M2**.
16. From Page Control, click **Design**, a dialog box will appear, select **M2**, and click **one arrow to the right**, then click **Design**.
17. Look at the results.
18. Create a proper report, and produce graphical output for both the Beam member, and the column member.

Module Review

1. Form Member is a function to create a continuous beam:
   a. True
   b. False

2. Concrete Design got two modes:
   a. Design Specification Mode, and Design Mode
   b. Design Creation Mode, and Design Layer Mode
   c. Design Layer Mode, and Design Mode
   d. None of the above

3. In the _____ we set the values of the design requirements.

4. Groups are:
   a. Beam Briefs + Column Briefs
   b. Briefs + Members
   c. Envelopes + Briefs
   d. Briefs + Job Info

5. In the Design Report I can show a detailed design results plus graphical illustration:
   a. True
   b. False

6. Use from the menus _____ to change the mode to Design Mode.

Module Review Answers

1. a
2. c
3. Briefs
4. b
5. a
6. Mode/Design

Module 10:

Steel Design

This module contains:

- Introduction to the new Steel Design
- Creating Envelopes, Members, Briefs, and Groups
- Check Code, and Member Select Commands
- The Design Report

Introduction

- Steel Design in STAAD used to be part of the input file of the Analysis, which means any change in the design commands means the input file should be re-run again, which is tedious and time consuming.
- Since STAAD.*Pro* 2003, a new separate-but-integrated Steel Designer module has been added, which reads the Analysis file, and perform design steel for selected members.
- Hence analysis is the first step of the Steel Design.
- From menus select Mode/Interactive Designs/Steel Design.
- STAAD.*Pro* will stay in the same interface (not like Concrete Design which takes the user to a separate program), but the menus, and the Page Control will change, as the below picture shows:

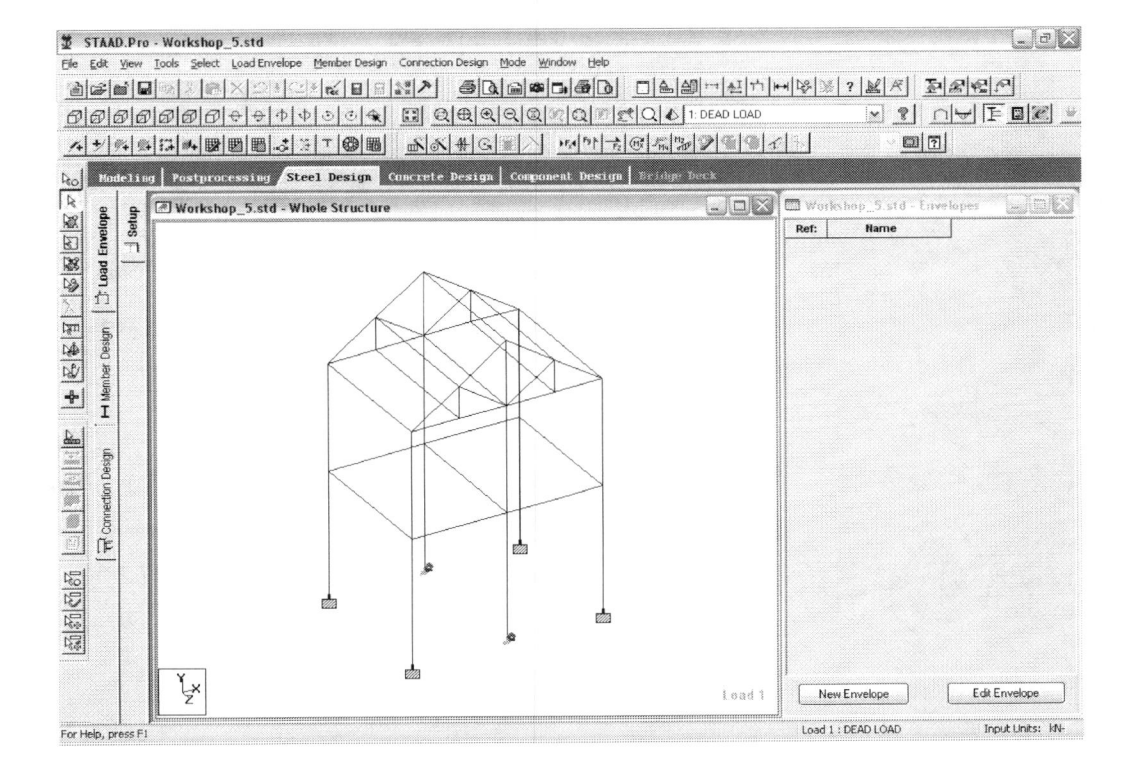

Step 1: Load Envelope Setup

- By default you are in **Load Envelope** page, and **Setup** sub-page.
- In Data Area, you can see two buttons:
  - New Envelope
  - Edit Envelope

[New Envelope]

- Click **New Envelope** to specify the Primary Loads and/or the Combinations, which you desire to be involved in the Design Process.

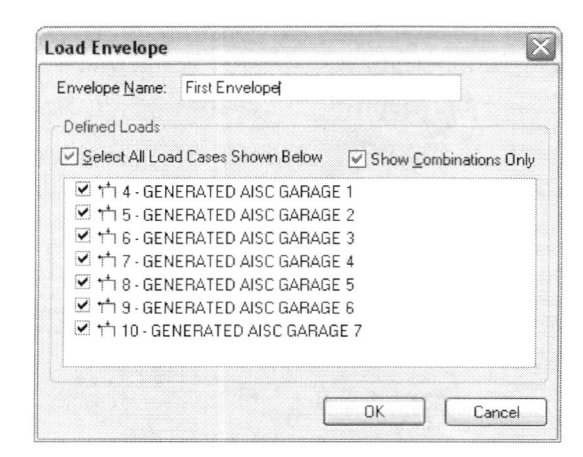

  - **Envelope Name**; input name for your envelope.
  - **Select All Load Cases Shown Below**; click this ON if all loads (namely Primary and Combination) are desired to be included in the envelope.
  - **Show Combinations Only**; if you want to show only the Combinations, and you are not interested in the Primary Loads.

- Once you are done click **OK**.

[Edit Envelope]

- Click **Edit Envelope**, if you want to edit an existing Envelope.

Step 2: Member Setup

- Click on the **Member Design** Page Control, the default sub-page will be **Member Setup**.
- This sub-page is to combine several adjacent beams into one single Member.
- Sometimes in the Modeling phase, user may insert intermediate Nodes inside a Beam for the sake of loading. As for STAAD.*Pro* in Steel design, it will treat any Beam between two Nodes as a separate Beam; hence a single Beam will not be considered deflecting together. In order to solve such a problem; combine Beams together.
- At the end, STAAD.*Pro* Steel Design will not design except Members, and not Beams and Columns coming from the Analysis file.
- To create Member, you have two ways:
  - Form Member
  - Auto Form Member

Form Member
- This function is usefeul to combine beams into single Member.
- Select the desired Beams you want to form as a single Member.
- From **Steel Design** toolbar click **Form Member** button.
- Automatically the selected Beams will form a single Member called M1 (if this is the first time you define Members in this file).

Auto Form Member
- This function is important to create non-stacked Column members, or single-non-adjacent Beams members.
- Select the desired Beams or Columns you want to form as a single Member.
- From **Steel Design** toolbar click **Auto Form Member** button.
- Automatically the selected Beams will form single Members and will be numbered as sequesnce M1, M2, M3, etc. or it will follow the last numbered Member.

- After creating Members, the screen may look something like:

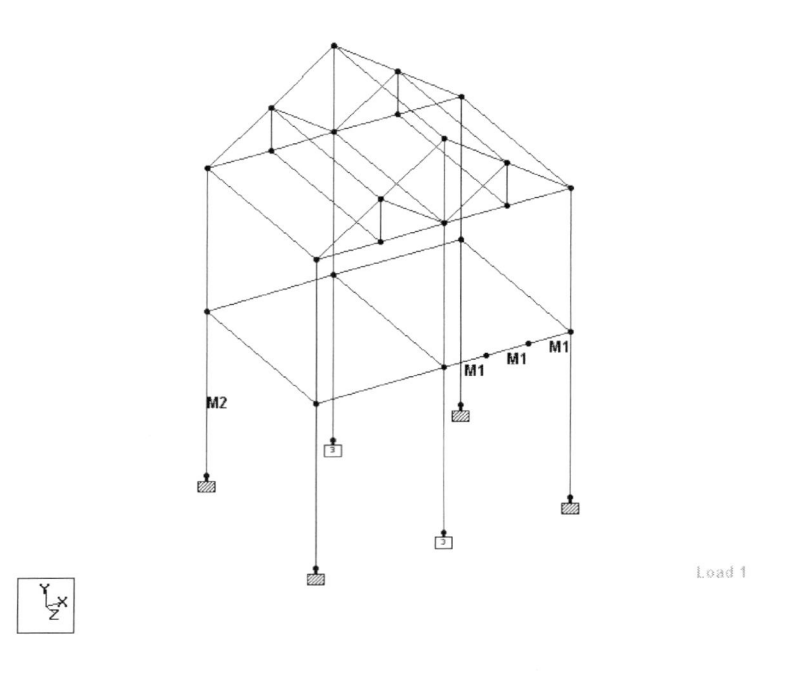

- Accordingly the Data Area would look something like:

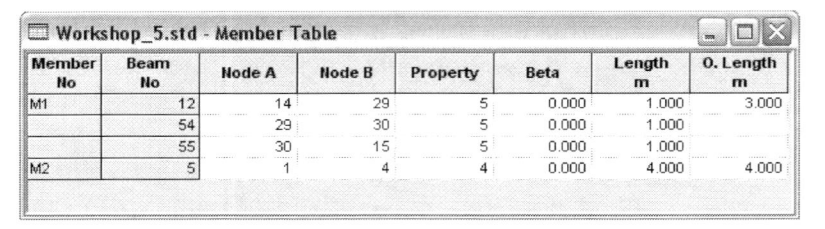

Step 3: Change the Restraints

- Click the **Member Design** Page Control, then sub-page **Restraints**.
- There are preset values for Member Restraints.
- In this sub-page you can change the values to reflect the real values of your case.
- Once you move to this sub-page, you will see the following table:

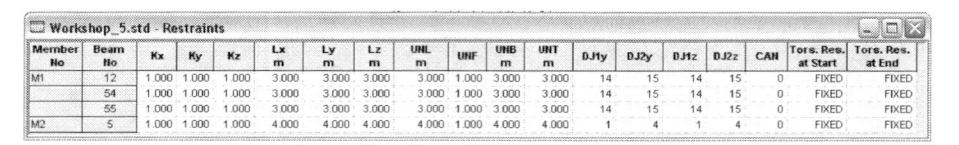

- To edit any of the values for any of the members:
  - From the table shown above simply select the desired member.
  - Right-click, the following shortcut menu will appear: 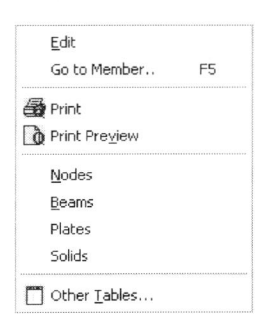
  - Click **Edit**, the following dialog box will appear: 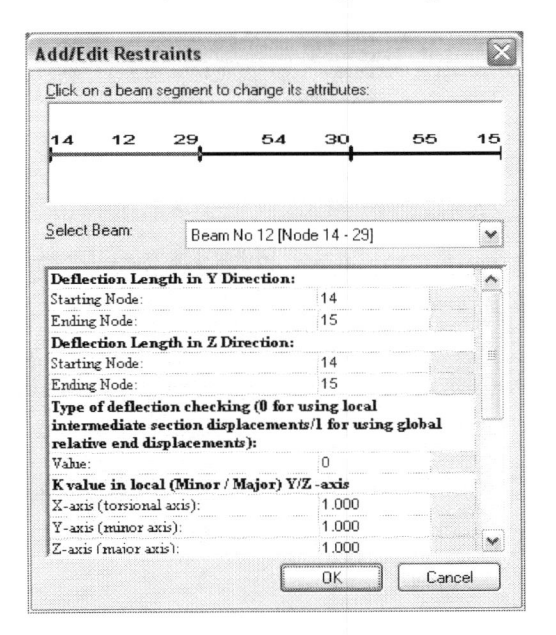
  - If the Member selected is consisting from more than one Beam, then STAAD.*Pro* will show all of the Beams, and you can select the desired Beam by two methods: either by clicking on the preview shape, or from the pop-up list.
  - Start changing the Restraint values as you wish, which they are:

Deflection length in Y Direction	■ To specify the Starting Node number and Ending Node number, so STAAD.*Pro* will calculate the defection in local Y direction for all the length between these two nodes. This is very useful if you (through the modeling phase) introduced intermediate nodes in a long beam. These are in the table DJ1y, and DJ2y.
Deflection length in Z Direction	■ The same thing applies for calculating the deflection in the local Z direction. These are in the table DJ1z, and DJ2z.
Type of deflection checking	■ Specify the Type of deflection, there are two possibilities: • Value of 0 (zero) means local intermediate section displacement. • Value of 1 means global relative end displacement.
K values	■ Slenderness factor in the three local axes x (torsional axis), y (minor axis), z (major axis).
Length	■ Effective Length in the three local axes x, y, z,
Unsupported Length for Allowable Bending Stress	■ The length of compression flange. ■ The same as above but as a factor, and not as a value. ■ The length of bottom flange ■ The length of top flange.
Tor. Res. At Start	■ To specify the boundary condition of the member for the purpose of designing for Torsion, they are: • **Free**: this means free end just like a cantilever. • **Pinned**: this means either a pinned support, or you set a Release using the **Specification** sub-page in the Modeling mode. • **Fixed**: This means either a fixed support, or a normal Node, which considered structurally a fixed end.
Tor. Res. At End	■ Just like above, except will be considered for the end Node.

Step 4: Creating Briefs

- In the **Briefs** we will set the values of the design requirements, which belong to this specific design case.
- The parameters you will control in the Brief will affect directly the design process. So a thorough study of these parameters means a better understanding of the design requirements.
- The Brief we will discuss is for **AISC ASD**. Click on the **Member Design** Page Control, and then sub-page **Briefs/Groups**.

- From the **Design Briefs** Data Area click **New Brief** button, the following dialog box will appear:

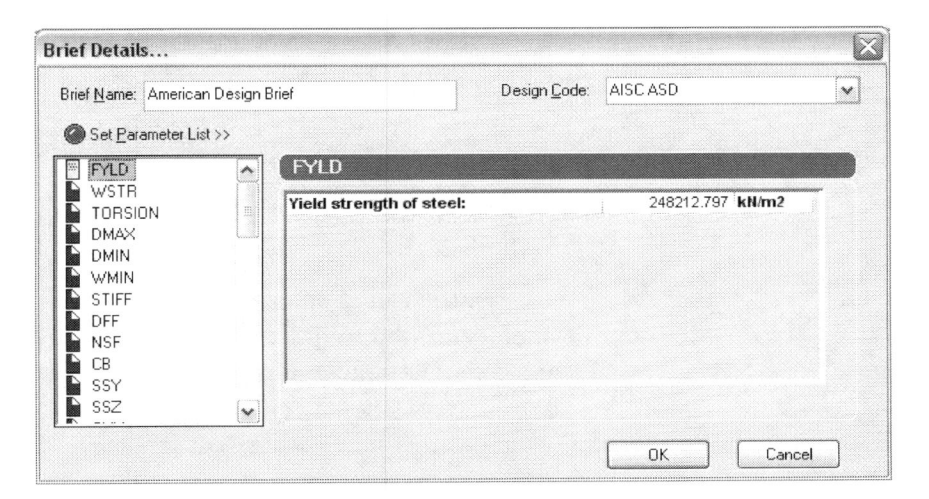

- Select the **Design Code** to be **AISC ASD**.
- Type a desired name for the Brief.
- Start changing the values, as your case needs.
- The parameters are:

Fyld ■ Yield strength of steel (*Default = 248212.797 KN/m2*)

Wstr ■ Allowable welding stress (*Default = 0.4 * Fyld*)

Torsion ■ Design for torsion, you have two choices:
- (0) Don't design for torsion (*Default*)
- (1) Design for torsion

Dmax, Dmin
- In the coming parts of this module we will discuss the **Member Select** command of STAAD.*Pro*, where the user will let STAAD.*Pro* select the proper cross section.
- STAAD.*Pro* will always select from the same cross section which the user specify in the Modeling mode and starting from the smallest cross section in the tables up to the largest cross section.
- You can limit the search by specifying **Dmin**, which is the minimum depth of the cross section and not the smallest cross section in the table. Also you can specify **Dmax**, which is the maximum depth of the cross section and not the largest cross section in the table.

Wmin
- Minimum welding thickness (*Default = 0.002m*)

Stiff
- Spacing for stiffeners for plate girder design

Dff
- No default value
- If you didn't input a value for this parameter, there will be no check for deflection.
- Dff is the Maximum Allowable Local Deflection.

Nsf
- Nsf is an abbreviation for **N**et **S**ection **F**actor.
  - Used to calculate A_{net}, which is $A_{net} = A_g \times Nsf$
  - By *default* the value is 1.00, which means the Gross Area equal to the Net Area.

Cb
- Cb as specified in chapter F of AISC ASD (*Default = 1*)

SSY, SSZ
- To specify if there will be any side sway in local y-axis and/or in local z-axis. You have two choices:
  - (0) There is sideway (Default)
  - (1) No sideway

CMY, CMZ
- CM value in local y-axis and in local z-axis. There are two choices:
  - SSY=0, hence there is sideway, therefore CMY = 0.85. Also, if SSZ=0, hence there is sideway, therefore CMZ = 0.85.
  - Or, if SSY=1, hence there is no sideway, therefore CMY will be calculated. Also, if SSZ=1, hence there is no sideway, therefore CMZ will be calculated.

Main
- Design for slenderness. You have two choices:
  - (0) Check for slenderness (*Default*)
  - (1) Don't check for slenderness

Ratio	■	Ratio of Actual Stresses to Allowable Stresses.
	•	By *default* this value is 1.00
	•	When design starts this Ratio will be governing the design, deciding whether certain cross section will SUCCEED or FAIL.
Weld	■	Design for weld. You have two choices:
	•	(1) Closed sections
	•	(2) Open sections (*Default*)
Beam	■	How to find the critical moments in designing a beam. You have two choices:
	•	Perform design at ends, and those locations specified in the SECTION command
	•	Perform design at ends, and 1/12th section location along member length (*Default*)
Profile	■	Only used for AISC ASD
	•	As discussed in Dmax, and Dmin, Profile will be also used in conjunction with **Select** command
	•	Profile will limit the selection process to a certain type (e.g. W4)
	•	You can select up to 3 profiles
Fu	■	Ultimate tensile strength of steel (*Default = 413688 KN/m^2*)
Taper	■	Decide how to design Taper I section, you have two choices:
	•	Design Tapered-I section based on rules of chapter F and appendix B of AISC ASD
	•	Design Tapered-I section based on appendix F of AISC ASD
Composite	■	Composite and all the below parameters will tackle the Composite Beam Design.
	■	Composite action with connectors. You have three choices:
	•	(0) No composite action (*Default*)
	•	(1) Composite act
	•	(2) Ignore positive moments during design
Condia	■	Diameter of shear connectors (*Default = 0.016m*)
Conheight	■	Height of shear connectors after welding (*Default = 0.064m*)
Cycles	■	Cycles of maximum stress that the shear connector is subjected to (*Default = 500000*)

Dlratio	■	Ratio of moment due to dead load applied before concrete hardens to the total moment (*Default = 0.4*)
Dlr2	■	Ratio of moment due to dead load applied after concrete hardens to the total moment (*Default = 0.4*)
Effwidth	■	Effective width of concrete slab (*Default = 0.25 Member Length*)
Fpc	■	Compressive strength of concrete at 28 days (*Default = 20684 KN/m^2*)
Impact	■	Impact fraction in % to be multiplied with live load (*Default = 20%*)
Plthick	■	Thickness of cover plate welded to the bottom flange of composite beam
Pltwidth	■	Width of cover plate welded to the bottom flange of composite beam
Ribheight	■	Height of rib from steel deck
Ribwidth	■	Width of rib from steel deck
Shoring	■	Temporary shoring during construction. You have two choices:

- (0) Without shoring (*Default*)
- (1) With shoring

Slabthick	■	Thickness of concrete slab, or thickness of concrete slab above the form steel deck (*Default = 0.102m*)
WMAX	■	Maximum weld thickness (*Default = 0.025m*)
FSS	■	Full Section Shear for welding, you have two choices:

- (0) False.
- (1) True. (*Default*).

OVR ■ In case of a member to accept or not an overstress:

- (1.00) means overstress is not permitted (*Default*).
- Any other value will be the factor which all of the allowable stresses multiplied by.

SHE ■ How the actual shear stress will be calculated:

- (0) means calculate Actual Shear stress using VQ/Ib (*Default*).
- (1) means calculate Actual Shear stress using V/(Ay or Az).

ELECTRODE ■ Specify the material used in welding from 6 different materials.

Module 10: Steel Design

DINC	■ This parameter and the parameters to follow are all for designing a Tapered section. DINC is depth of tapered section to be incremented.
FTBINC	■ Flange width (top and bottom) of tapered section to be incremented.
FTINC	■ Top flange to be incremented only. In this case Bottom Flange = W.
FBINC	■ Bottom flange to be incremented only. In this case Top Flange = W.
BMAX	■ Maximum allowable width of the flange.
Set Parameter List >>	■ Use this button in the **Brief Details** dialog box if you want to minimize the parameters you want to set.
	■ If you click it, the following will be shown:

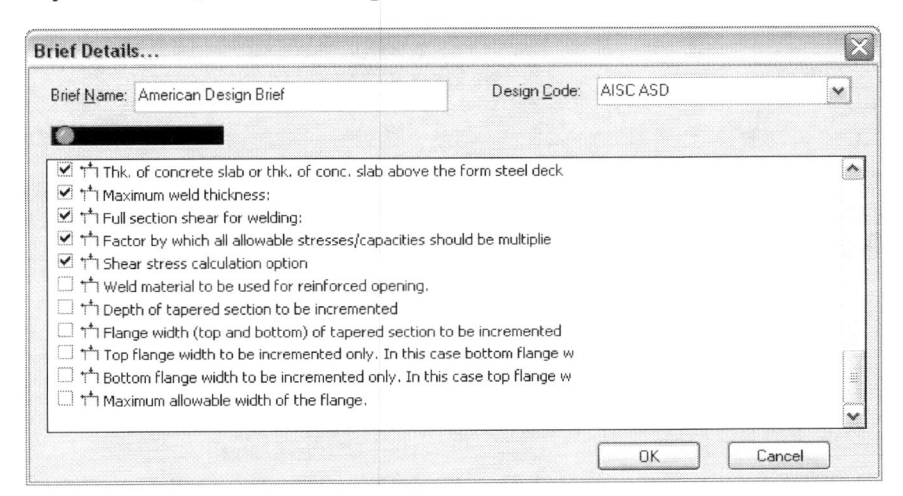

	■ Click OFF the unneeded parameters, hence only the selected parameters will be available for editing.
Note	■ In order to control these parameters, user should refer always to STAAD manuals as it refers to the chapters, clauses, and parameters in AISC-ASD, which will enable the user to relate each STAAD parameter to the correct required part of the code.

10-13

Step 5: Creating Design Groups

- To combine: Envelope, Member, and Brief
- These are the steps to create a **Design Group**:
  - Make sure you are in the **Briefs/Groups** sub-page, if not, go to it.
  - Select the desired Member to design.
  - Click the **New Design Grp** (Grp means Group) button.
  - The following dialog box will appear:

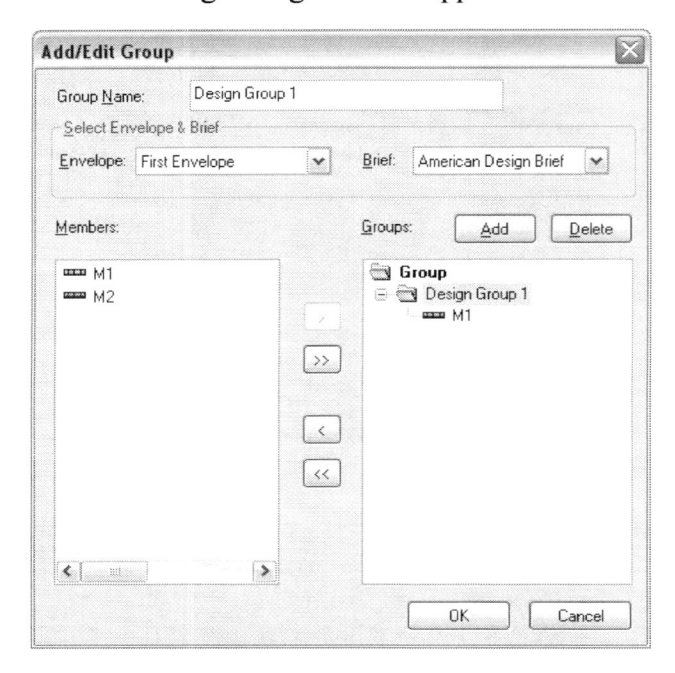

- Type in the **Group Name**.
- From the pop-up lists select an **Envelope**, and then select **Brief**.
- Click **Add** button, a new Group will be added.

- Because the desired Member is already selected, simply click one arrow to the right button, so the Member will be associated with the Group. When you are done click **OK**.

Module 10: Steel Design

- You can see the results of what you did in Data Area in the **Design Groups** table, just like the following:

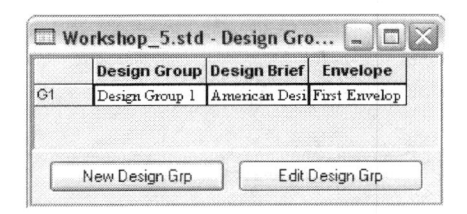

- You will see the Design Group name, the Design Brief name, and finally the Envelope name.
- To edit any part of the Group, simply select it an click **Edit Design Grp** button.

Steel Design Commands in STAAD.*Pro*

- After you finish the input, execute one of two steel design commands in STAAD.*Pro*:
  - Check Code
  - Member Selection

Check Code
- STAAD.*Pro* will take the cross-section selected by the user and check it against the loads (different types of stresses), and report whether the cross-section is SAFE, or UNSAFE.
- The Brief selected has a major effect over the results.
- As noticed before that, the parameters will control which way should STAAD.*Pro* take in order to decide the outcome of the design process.

Member Selection
- STAAD.*Pro* will select from the table the lightest section, which will sustain the loads input by the user.
- STAAD.*Pro* will select the same type of the cross-section the user designated in the first place.
- Also, the Brief (parameters) input by the user, will decide which way STAAD.*Pro* will take in order to select the suitable cross-section.

Command Execution
- Make sure that the desired Member to be designed is selected.
- From menus select Member Design/Perform Group Design, and then select either Check Code, or Member Selection.
- The following dialog box will appear:

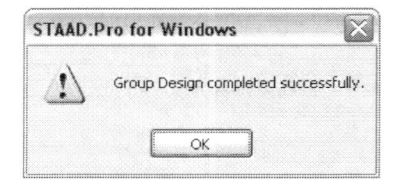

> STAAD.Pro for Windows
> Group Design completed successfully.
> OK

- Click **OK**.
- In order to see the results, simply go the sub-page titled **Results/Reports**, in the lower part of the screen. You will see the following table:

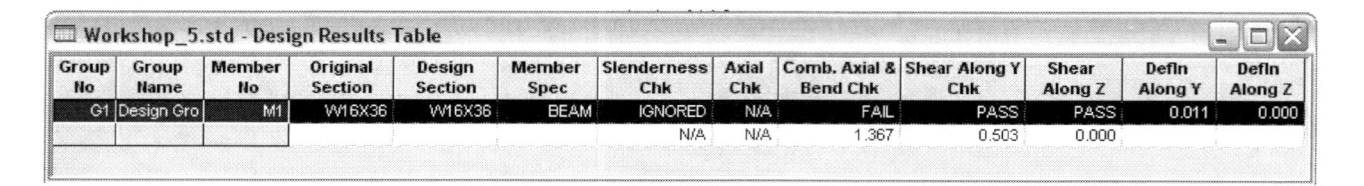

Group No	Group Name	Member No	Original Section	Design Section	Member Spec	Slenderness Chk	Axial Chk	Comb. Axial & Bend Chk	Shear Along Y Chk	Shear Along Z	Defln Along Y	Defln Along Z
G1	Design Gro	M1	W16X36	W16X36	BEAM	IGNORED	N/A	FAIL	PASS	PASS	0.011	0.000
						N/A	N/A	1.367	0.503	0.000		

- In order to see a full report, do the following steps:
  - Select the **Group No.** (in the above illustration the Group No. is **G1**)
  - Right-click, shortcut menu will appear select from it the option **View Design Calculation**.
  - You will see something like the report in the next page:

10-16

Module 10: Steel Design

Design Calculation Report

- The following is Calculation Report:

Design Of Member No. 1 As Per AISC

Input Parameters

Member Section	W16X36
Cross Sectional Area A_x (m^2)	0.01
Shear Area Along Z Axis A_z (m^2)	4.38
Shear Area Along Y Axis A_y (m^2)	4.20
r_z (m)	0.17
r_y (m)	0.04
Section Modulus About Z Axis - Tension Edge S_{tz} (m^3)	0.00
Section Modulus About Z Axis - Compression Edge S_{zz} (m^3)	0.00
Section Modulus About Y Axis - Tension Edge S_{ty} (m^3)	0.00
Section Modulus About Y Axis - Compression Edge S_{yy} (m^3)	0.00
Unsupported Length - Z Axis For Slenderness Check L_z (m)	3.00
Unsupported Length - Y Axis For Slenderness Check L_y (m)	3.00
Effective Length For Allowable Bending Stress Calculation Unl (m)	3.00
Yield Stress f_y (MPa)	248.00
Ultimate tensile strength f_u (MPa)	414.00
Allowable Ratio For Interaction Check	1.00

Design Forces

Combined Axial Force & Bi-axial moment

Axial Load F_x (kN)	42.87

Printing the Report

- While the report is open, select from menus **File/Print Preview**, and from the Print Preview window you can direct the output to the printer.

- Or you can reach to the same results by using the **Print** toolbar; click the **Print Preview Report** icon.

Steel Design

Workshop 7-B

1. Open Small_Building file.
2. Select **Mode/Interactive Designs/Steel Design**.
3. Define new Envelope call it **First Envelope** to contain all the loads (Primary and Combination).
4. Create the following members as shown:

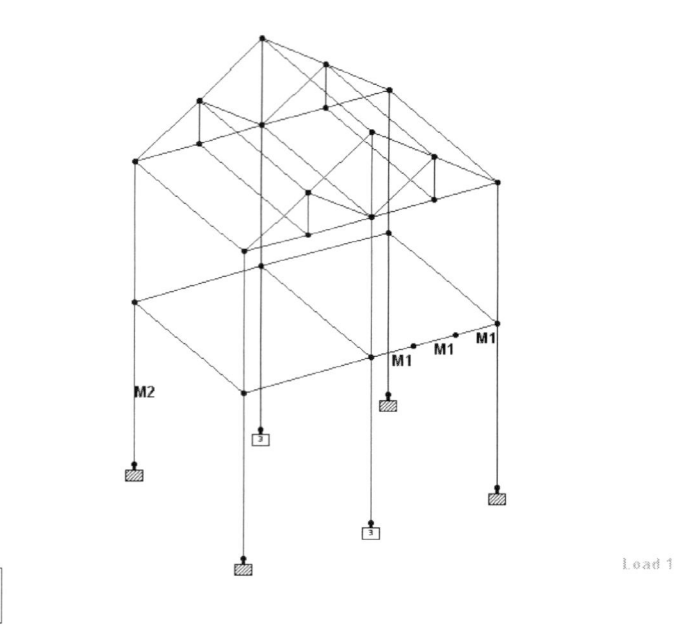

5. For M1, remember the leftmost Node number, and the rightmost Node number, now go to Restraint sub-page and check the numbers of DJ1y, and DJ2y, are they the same numbers? Or not? If not change them to the Node numbers you know.

6. Create a new Brief, based on AISC-ASD, and call it *First Brief*, and change the following parameters:

    a. Torsion = 1

    b. Dff = 180

    c. Nsf = 0.85

    d. Ssz = 1

    e. Main = 1

    f. Ratio = 0.95

    g. Beam = 1

7. Select M1

8. Create a new Group and call it *First Group*, and combine:

    a. First Group

    b. First Brief

    c. First Envelope

    d. M1

9. Select from menus **Member Design/Perform Group Design/ Check Code**.

10. From Design Calculation report, did STAAD.*Pro* check for Slenderness? Why?

11. What was the critical stress? _____, And what was the ratio value? _____

12. What was the actual deflection in the local y-axis, and what was the allowable deflection?

13. Create another Brief, based on AISC-ASD, and call it *Second Brief*, and change the following parameters:
    a. Dmax = 1m
    b. Dmin = 0.3m
    c. Nsf = 0.85
    d. Ssz = 1
    e. Main = 0
    f. Ratio = 0.95
    g. Beam = 0
14. Select M2
15. Click **Edit Design Grp**, and delete First Group
16. Create a new Group and call it *Second Group*, and combine:
    a. Second Group
    b. Second Brief
    c. First Envelope
    d. M2
17. Select from menus **Member Design/Perform Group Design/ Member Selection**.
18. From Design Calculation report, if you know that the assumed cross-section was HP14X102, what is the selected cross-section of STAAD.Pro _____.
19. Did STAAD.*Pro* check for deflection? Why?
20. What was the critical stress? _____, and what was the ratio value? _____.

Module Review

1. The new Steel Design Engine is separate-but-integrated engine:
    a. True
    b. False

2. There are two steel design commands, which they are:
    a. Beam Design and Column Design
    b. Check Code and Beam Design
    c. Member Select and Column Design
    d. Check Code and Member Select

3. Design Group combines _____, _____, and _____.

4. In the Brief you can specify:
    a. To check for Slenderness
    b. To check for Deflection
    c. The maximum depth of the cross-section if STAAD will perform Member Select
    d. All of the above

5. Envelopes can be created only from Combinations:
    a. True
    b. False

6. You can change Kx, Ky, and Kz in _____ sub-page.

Module Review Answers

1. a
2. d
3. Envelopes, Brief, and Member
4. d
5. b
6. Restraints

Test Your Knowledge

The main objective of this test is to test-your-knowledge, which you learned from this tutorial. The test should be closed book, and without using the software.

How to assess your self:

- First don't look at the answers in the last page
- If you score 25 or more out of 30, then you are Excellent
- If you score between 20-24, then you are Very good
- If you score between 15-19, then you are Good
- If you score below 15, then you are Fair

STAAD.Pro 2005 Tutorial

Name:_____   Date:_____

Answer *All* of the following questions:

1. Geometry in STAAD.*Pro* is:
    a. Nodes
    b. Beams
    c. Plates
    d. None of the above
2. You create Combinations, then you create Primary Loads:
    a. True
    b. False
3. Assigning Truss Members to certain Beams can be done in:
    a. Properties
    b. Specs
    c. Loading
    d. Analysis
4. How many types of Static Analysis are there in STAAD.*Pro*:
    a. 1
    b. 2
    c. 4
    d. 3
5. Which of the following is *not* considered a method to create geometry in STAAD.*Pro*:
    a. Drafting the Geometry
    b. Importing DXF file
    c. Importing DOC file
    d. Importing XLS file
6. You have to select Beams prior to creating the Floor Load:
    a. True
    b. False
7. How many files STAAD.*Pro* can deal with simultaneously:
    a. 1
    b. 2
    c. 3
    d. 4
8. Which is true about selecting methods in STAAD.*Pro*
    a. Window
    b. Clicking
    c. Clicking the table
    d. All of the above
9. One of the following specification does not apply on L-sections:
    a. LD
    b. TB
    c. SD
    d. RA

10. Beta can be:
    a. +90
    b. –90
    c. +90 and –90
    d. Any angle
11. In Post Processing user can generate a handsome reports, with different fonts, embedded graphics, showing the logo of the company:
    a. True
    b. False
12. In the new Concrete Design module:
    a. There will be new menus
    b. There will be new Page Controls
    c. There will be new Modes
    d. All of the above
13. Node in the input file is described by
    a. The Beam at its ends
    b. Node Number
    c. Node Coordinates
    d. B & C
14. After you assign a cross-section to Beam, then double-clicking it, one of the following will not be shown:
    a. Cross-sectional area
    b. Volume
    c. Moments of Inertia
    d. Effective Shear Areas
15. The title of the Combination is very important to STAAD.*Pro*, and can predict from it the factors to be multiplied by the Primary loads.
    a. True
    b. False
16. One of the results STAAD.*Pro* will generate is the rotational of Nodes around X, Y, Z, the units of this piece of information is:
    a. Degrees
    b. Radians
    c. Grads
    d. User choice
17. In Translational Repeat command, the direction of repeat is:
    a. In the Global direction only
    b. In the Global and Local direction
    c. In the Local direction only
    d. Non of the above
18. The Minor principle axis is Local Z
    a. True
    b. False

19. Fixed But support will allow the user to release one or more of the six reactions of the that support:
    a. True
    b. False
20. In Steel Design module Ky, Kx, Ly, UNL, DJy1, these are:
    a. Briefs
    b. Restraints
    c. Deflection Length parameters
    d. Member Controlling specifications
21. The Analysis type, which takes care of, both secondary loading caused by Displacement of the Nodes, and geometric stiffness correction caused by Deflection of Beams is:
    a. Perform Analysis
    b. P-Delta Analysis
    c. Non-Linear Analysis
    d. Time History Analysis
22. The file extension of STAAD.Pro input files is:
    a. *filename*.pro
    b. *filename*.stp
    c. *filename*.std
    d. *filename*.sta
23. Although we create our input file using graphical methods, STAAD.Pro will create a text file which represent the input file containing the STAAD.Pro syntax:
    a. True
    b. False
24. The Material constants stored in STAAD.Pro are:
    a. The only method to input Material constants for your Beams
    b. You can input your own if you didn't like them
    c. Can be altered
    d. There are no stored Material constants in STAAD.Pro
25. Bending Moment in STAAD.Pro is the moment:
    a. Around y-y
    b. Around x-x
    c. Around z-z
    d. Around Global Z
26. In Circular Repeat, the resulting total number of frames is:
    a. No. of Steps assigned by user + 1
    b. No. of Steps assigned by user − 1
    c. No. of Steps assigned by user
    d. No. of Steps assigned by user + 2
27. In Concrete Design/Beam Briefs, where do you link Envelope to Brief:
    a. In the Brief dialog box, General tab
    b. In the Brief dialog box, Column Factors tab
    c. In the Brief dialog box, Main Reinforcement tab
    d. In the Group dialog box
28. In Concrete Design/Column Brief, where do you link Envelope to Brief:
    a. In the Brief dialog box, General tab
    b. Column Briefs uses Combinations rather than Envelopes
    c. In the Brief dialog box, Column Factors tab
    d. In the Group dialog box
29. In Post Processing Mode:
    a. I can change the Scale of the results
    b. I can change the Units of the results
    c. I can show or hide the values of the results
    d. All of the above
30. In Rotate Command, I can specify the Axis of rotation by:
    a. Typing in the numbers of two Nodes
    b. Clicking on two Nodes through a special cursor
    c. Type in the coordinates of two Nodes
    d. All of the above

Answers:

1. a
2. b
3. b
4. d
5. c
6. b
7. a
8. d
9. b
10. d
11. a
12. d
13. d
14. b
15. b
16. b
17. a
18. b
19. a
20. b
21. c
22. c
23. a
24. b
25. c
26. a
27. a
28. b
29. d
30. d